Aviation and Sustainability

This book offers a comprehensive and management-oriented analysis of sustainability in the aviation industry. It examines aviation's crucial role in global mobility and economic development whilst also critically addressing its environmental, social and regulatory challenges. Readers will benefit by gaining a realistic, analytically sound framework for evaluating sustainability strategies in aviation. The book supports informed decision-making, helps to anticipate regulatory and market developments, and enables readers to better balance economic performance with environmental responsibility and social acceptance in a highly complex and globally interconnected industry.

By adopting a holistic ESG perspective, the book integrates environmental impacts such as climate change and noise, social issues including equity, accessibility and labour conditions, and governance aspects like regulation, reporting and stakeholder accountability. Rather than presenting simplified solutions, it highlights trade-offs, uncertainties and long-term path dependencies that shape strategic decision-making in aviation. The structured approach combines conceptual foundations with industry-specific insights, covering regulation, technology, sustainability reporting and strategic management practices.

The book is relevant for all managers in the aviation industry, whether they work for airlines, airports, air traffic control organisations, regulatory authorities or political decision-makers. It is particularly important for environmental managers throughout the aviation industry, but also for managers at a senior level, as sustainability is becoming a strategic issue of growing importance. In addition, the book is valuable for students in aviation management programmes as well as for students in related fields such as aeronautical engineering and tourism management.

Karsten Benz (Prof. Dr.) is Professor of Air Traffic Management at Worms University of Applied Sciences. His research focuses on ATM/UTM and Sustainable Aviation Fuel. After 25 years on C-Level at Lufthansa Group, Karsten Benz works now as Independent Advisor and Board Member.

Frank Fichert (Prof. Dr.) is Professor of Economics and Transport Economics at Worms University of Applied Sciences. His research focuses on competition and regulation in the air transport industry and the environmental issues of aviation.

Tobias Grosche (Prof. Dr.) is Professor of Aviation Management at Worms University of Applied Sciences. His teaching and research focus on network and flight planning, flight operations, aviation analytics and sustainability in aviation.

Managing Aviation Operations

Series Editor: Barry Humphreys

The purpose of this series is to provide a comprehensive set of materials dealing with the key components of airline and airport operations. To date, this innovative approach has not been evident among aviation topics and certainly not applied to operational areas of airlines or airports. While more recent works have begun, in brief, to consider the various characteristics of operational areas, the *Managing Aviation Operations* series will expand coverage with far greater breadth and depth of content.

Airlines and airports are devoid of specific topic knowledge in ready-made, easy-to-read, creditable resources. Tapping into industry expertise to drive a range of key niche products will resource the industry in a way not yet seen in this domain. Therefore, the objective is to deliver a collection of specialised, internationally sourced and expertly written books to serve as readily accessible guides and references primarily for professionals within the industry. The focus of the series editors will be to ensure product quality, user readability and appeal, and transparent consistency across the range.

Operational Safety for Aviation Managers
Practice Beyond the Theory
John Frearson

Innovative Risk Leadership in Aviation
Industry Insights and Cases
Triant Flouris, Ayşe Küçük Yılmaz and Konstantinos Malagas

Airport Management
Investing in, Operating and Growing the Business
Tan Sri Dr Bashir Ahmad Bin Abdul Majid and Edward Clayton

Aviation and Sustainability
Management, Technologies and Accountability
Karsten Benz, Frank Fichert and Tobias Grosche

For more information about this series, please visit: https://www.routledge.com/Managing-Aviation-Operations/book-series/MAO

Aviation and Sustainability

Management, Technologies and Accountability

Karsten Benz, Frank Fichert and Tobias Grosche

Routledge
Taylor & Francis Group
LONDON AND NEW YORK

Designed cover image: Getty Images © Credit: Rob Wilkinson

First published 2027
by Routledge
4 Park Square, Milton Park, Abingdon, Oxon OX14 4RN

and by Routledge
605 Third Avenue, New York, NY 10158

Routledge is an imprint of the Taylor & Francis Group, an informa business

British Library Cataloguing-in-Publication Data
A catalogue record for this book is available from the British Library

ISBN: 978-1-041-00104-1 (hbk)
ISBN: 978-1-041-00090-7 (pbk)
ISBN: 978-1-003-60817-2 (ebk)

DOI: 10.4324/9781003608172

Typeset in Optima
by Taylor & Francis Books

Contents

Illustrations

Figures

Tables

Boxes

1 Introduction

Mobility is a fundamental prerequisite for economic activity, social interaction and cultural exchange. In this context, aviation has become a decisive factor for global connectivity. Air transport facilitates international trade, tourism, business relations and interpersonal exchange. It does so on a scale and at a speed that no other mode of transport can match. For many remote regions, including a large number of islands, aviation is not just an alternative mode of transport, but an essential prerequisite for economic development and social participation. In recent decades, flying has evolved from a luxury good to a widespread form of everyday mobility, especially in North America and Europe but also more and more in Asia, and has thus contributed significantly to the emergence of a mobile and highly globalised society.

At the same time, the aviation sector is increasingly the focus of public and political debate due to its environmental impacts and social aspects. Although the economic and social benefits of air transport are considerable, they are accompanied by growing environmental, social and distributional challenges. In particular, the contribution of air transport to climate change has come under criticism, as the environmental impacts associated with air transport are largely externalised and borne by society and natural ecosystems rather than by the industry itself. Beyond environmental impacts, social aspects such as fairness, public acceptance and corporate responsibility are becoming increasingly important. This changing context is fundamentally altering the operating environment for airlines and increasing the pressure for structural transformation.

This constellation leads to a fundamental paradox. On the one hand, global demand for air travel is expected to rise further in the coming decades, mainly due to rising incomes, a growing middle class in emerging economies and the continuing importance of global mobility for economic integration. On the other hand, this growth trajectory is increasingly coming into conflict with stricter climate targets, an expanding regulatory framework and changing social expectations, particularly in industrialised countries. From this perspective, the term "sustainable aviation" may seem contradictory at first glance, as continuous demand growth appears difficult to reconcile with ambitious environmental goals.

DOI: 10.4324/9781003608172-1

The challenge is amplified by the structural attributes of the aviation industry. Aviation is characterised by high capital intensity, long investment and asset life cycles, strict regulatory oversight and deep global interdependencies across value chains and operating networks. Aircraft, infrastructure and related technologies require significant upfront investment and are used over several decades, limiting short-term flexibility and slowing the pace of transformation. Furthermore, aviation operates in a highly standardised and safety-critical environment where changes must be carefully coordinated and internationally harmonised. As a result, strategic decisions in aviation tend to have long-term consequences, while management's room for manoeuvre is limited when compared to less regulated industries.

Regulation plays a central role in shaping sustainability outcomes in aviation. Since the early days of civil aviation, the sector has relied heavily on international standardisation to ensure safety, interoperability and economic efficiency. This has enabled the global expansion of air transport. Any deviation in the form of regional rules poses a challenge for corporate governance. Environmental and reporting regulations are increasingly emerging at various levels, globally, internationally and nationally, often with different scopes, objectives and implementation mechanisms. This fragmentation increases complexity for globally active companies and raises concerns about competitive distortions and carbon leakage, where emissions are shifted geographically rather than reduced in absolute terms. Consequently, regulation is both an important driver of the transformation towards sustainability and a source of strategic uncertainty for managers in the aviation industry.

Against this backdrop, sustainability has evolved from a marginal issue or a matter of compliance to a key strategic concern for aviation management. The long-term profitability of companies increasingly depends on their ability to respond proactively to environmental and social challenges while maintaining their economic performance. It is important to note that there is no single technological or managerial "silver bullet" that can ensure sustainable aviation. Technological innovations – such as more fuel-efficient aircraft, operational improvements and the development of Sustainable Aviation Fuel (SAF) – play a crucial role but are not sufficient on their own. Similarly, a one-sided focus on regulatory compliance fails to take into account the broader strategic and societal dimensions of sustainability.

This book takes a holistic approach to sustainability, which is conceived as a multidimensional construct encompassing environmental, social and economic dimensions, often summarised under the ESG framework (environmental, social and governance). Environmental aspects include climate and local environmental impact, while social dimensions address issues such as equity, accessibility and social acceptance. Governance refers to the institutional framework that shapes the setting, implementation and monitoring of sustainability goals, including transparency, accountability and stakeholder participation. However, economic viability remains a necessary prerequisite for any meaningful transformation, given the scale of investment required in new aircraft, infrastructure and alternative energy sources. The economic growth enabled by aviation can support broader development goals, but it can also intensify social tensions if costs and benefits are unevenly distributed across regions and stakeholder groups.

The aim of this book is to offer an integrated, management-oriented perspective on sustainability in aviation. Rather than offering simplified solutions, it provides managers, policymakers and practitioners with analytically sound insights into the complex trade-offs, uncertainties and path dependencies that characterise sustainability-related decisions in the aviation sector. By explicitly considering institutional, regulatory, technological and market-related constraints, the book aims to contribute to a more realistic and actionable understanding of sustainable transformation in aviation.

The structure of the publication reflects this integrative approach. Following this introductory chapter, the subsequent chapters deal with the conceptual foundations of sustainability, the economic and social contributions of aviation, its environmental impacts and the changing regulatory environment. Further chapters examine technological and operational measures, sustainability reporting practices, strategic management approaches, and sustainability-oriented marketing and communication. The book concludes with summarised, stakeholder-specific recommendations designed to support informed decision-making across the aviation ecosystem.

By combining conceptual analysis with industry-specific insights and a corporate-level perspective, this book contributes to an informed, practice-oriented debate on how aviation can navigate the tension between its undeniable societal value and the urgent need for greater sustainability.

2 Aviation in the Context of Sustainability

Introduction

Aviation plays a critical role in global mobility and economic development, yet it is increasingly challenged by rising expectations regarding environmental and social responsibility. As air traffic continues to grow, the sector's contribution to climate change, combined with its reliance on fossil fuels and long investment cycles, complicates the transition towards sustainability. At the same time, fragmented regulatory frameworks, shifting stakeholder demands and concerns about social equity place additional pressure on industry actors. This chapter situates aviation within the broader sustainability debate and examines the industry's structural characteristics, regulatory context and stakeholder dynamics. By adopting a multidimensional perspective, it highlights why sustainable transformation in aviation is both essential and inherently complex.

Sustainability as a Multidimensional Concept

Since its original use in forestry in the 18th century, the concept of sustainability has developed into a central guiding principle of global politics and economics.[1] The definition that is commonly used today goes back to the Brundtland Report of the World Commission on Environment and Development, which describes sustainability as "development that meets the needs of the present without compromising the ability of future generations to meet their own needs".[2] The concept of the triple bottom line, which understands sustainability as the interplay of ecological, social and economic goals, has become established in the economic debate. These three dimensions are often referred to as "people, planet, profit" and form the basis for corporate sustainability strategies (see Figure 2.1).[3]

In corporate practice, sustainability is increasingly being operationalised through ESG (Environmental, Social, Governance) criteria. These criteria serve investors and stakeholders as a benchmark for sustainable business practices.[4] At the same time, the United Nations' 17 Sustainable Development Goals (SDGs) have created a global orientation framework that is also relevant for the aviation industry.[5]

DOI: 10.4324/9781003608172-2

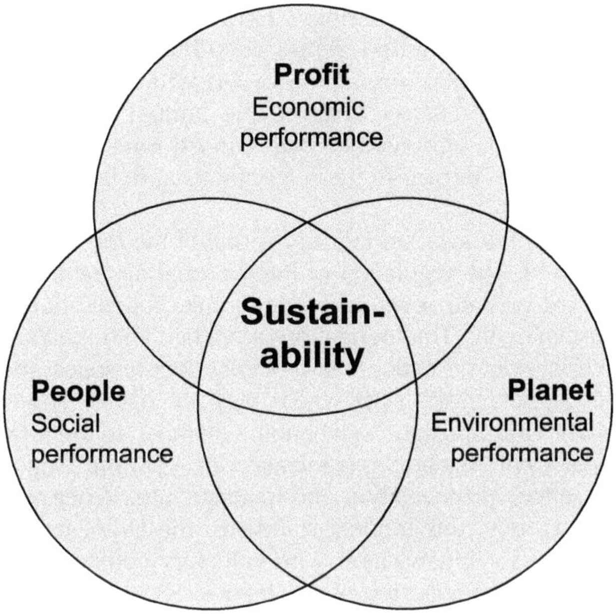

Figure 2.1 Multidimensional concept of sustainability in aviation.

Another key concept is Corporate Social Responsibility (CSR), i.e., the voluntary social responsibility of companies that goes beyond legal requirements. CSR encompasses measures in the areas of the environment, working conditions, human rights and a focus on public welfare. In aviation, for example, CSR includes noise reduction programmes and regional development partnerships around airports.[6]

Characteristics of the Aviation Industry

Over the past three decades, global passenger air traffic has consistently outpaced economic growth, exceeding global GDP expansion more than twice.[7] However, this trend has slowed in recent years. For the period 2024–2026, demand growth is expected to align more closely with GDP and be about 1.5 times higher. The International Air Transport Association (IATA) reported an increase in demand of 5.3% year-over-year by the end of October 2025, with an average load factor of 83.7% forecast for 2025. More moderate growth is expected for 2026, although this growth varies widely within different regions of the world. While Europe and North America expect only slight growth, the Asia-Pacific region, Africa and the Middle East are developing strongly. Technological innovations are increasingly characterising the industry. AI-supported systems, biometrics, predictive maintenance, digital twins and optimised flight routes promise significant cost savings and attractive customer service offerings. However, the industry is facing structural challenges. Production delays at Airbus and Boeing are leading to backlogs of over 17,000 aircraft on order, which could take well into the 2030s to be delivered. As a result, the annual improvement in fuel efficiency of 1.5 to 2% achieved to date can no longer be maintained. In addition, geopolitical tensions, trade conflicts and labour shortages are slowing down growth.[8]

At the same time, the aviation industry is one of the most heavily regulated sectors in the world. The regulation of international air traffic has developed within the context of state sovereignty, economic liberalisation and growing pressure for sustainability. The foundations were laid with the Paris Agreement of 1919, which introduced national sovereignty over airspace and the obligation to register aircraft in the state where they are based. The breakthrough came in 1944 with the Chicago Convention, which led to the establishment of the International Civil Aviation Organization (ICAO) and a global standard framework for safety, air navigation and infrastructure. While the ICAO, as a United Nations agency, sets binding standards, the IATA, founded in 1945, acts as a private-sector organisation with voluntary commitments and advocates for the global harmonisation of regulations. A paradigm shift took place in the 1970s and 1980s as a result of liberalisation and deregulation policies such as the Airline Deregulation Act in the USA in 1978 and the creation of the European internal aviation market from 1992. The development of the European Union (EU) aviation market towards a liberalised "open skies" market took place in several phases and had far-reaching implications for competition, prices, route networks and corporate structures. Until the 1980s, European air traffic was strongly regulated on a bilateral basis. Flight permits were granted to national flag carriers (e.g., Lufthansa, Air France) and capacities, prices and frequencies were negotiated between governments. In 1987, first steps towards liberalisation were taken, including the introduction of multiple traffic rights on routes. The third liberalisation package from 1992 had the greatest impact with the creation of a single aviation market. Starting in 1993, all EU airlines gained the right to operate on any intra-EU route without capacity and price regulation and on the basis of uniform operating licences and safety standards.[9] As a result, there was a rapid growth in low-cost carriers, combined with falling ticket prices and a widening range of flights. From 2002, successive open skies agreements were concluded with third countries, e.g., the EU-US Open Skies Agreement between the EU and the USA. At the same time, however, new requirements for market regulation and consumer protection emerged (e.g., EU Regulation 261/2004 on air passenger rights).

The long-term economic development of the industry depends crucially on how efficiently technological innovations are implemented and how effectively the regulatory framework supports the transformation, particularly with regard to sustainability. The aviation industry is characterised by several specific attributes in the context of sustainability:

- High Capital Intensity: Aircraft have a lifespan of around 25–30 years. Investment decisions are correspondingly long term. The industry only reacts to technological innovations over a longer period of time, as safety requirements, certification deadlines, high investment necessities and lengthy renewal of aircraft fleets slow down adaptation.
- High Climate Impact: In addition to CO_2 emissions, which are low compared to many other industries, aviation's non-CO_2 effects are also significant, caused by contrails and NO_x emissions (see Chapter 4). Due to these non-CO_2 effects aviation's climate impact is 2–3 times higher than its share in CO_2 emissions.
- International Regulation: Aviation is only partially included in the international framework for climate protection. While domestic air traffic is considered part of national emissions balances and is therefore covered by the Paris Climate Agreement, international air traffic is explicitly excluded from its scope. This institutional separation is a special aspect of international climate policy and makes it difficult to coherently integrate aviation into global emission reduction strategies. Internationally coordinated regulations, e.g., via the ICAO, are essential.
- Dependence on Fossil Energy: While electromobility is developing fast in the automotive industry, the energy efficiency of batteries is currently insufficient for air traffic, except for small regional jets. Alternatives such as Sustainable Aviation Fuel (SAF) are feasible from a technological perspective but are not yet available on a scalable basis and are significantly more expensive than jet fuel (see Chapter 6). Aviation is therefore difficult to decarbonise.
- Dynamics of Growth: In aviation, enhanced engine performance and aerodynamic design optimisation have significantly reduced fuel consumption per passenger and tonne-kilometre. However, these efficiency gains have contributed to lower operating costs, which in turn have made air transport more affordable. This has contributed to the sharply rising demand for air services, leading to a net increase in overall fuel consumption and associated emissions. This so-called rebound effect emphasises the limits of relying solely on efficiency improvements to mitigate climate change.[10] According to forecasts, global passenger volumes will more than double by 2050, which will intensify the challenge of decarbonisation.[11] Aviation's share of global CO_2 emissions could rise to 12–18% by 2050 if no effective countermeasures are taken. This development is primarily due to the above-average growth dynamics of the aviation sector and the slower rate of decarbonisation compared to other industries. While electrification and alternative energy sources are increasingly leading to falling emissions in sectors such as electricity generation and ground transport, aviation remains heavily reliant on fossil fuel. Technological options such as SAF are still at an early stage of development and market diffusion and are currently neither economically nor infrastructurally scalable. At the same time,

annual operational efficiency increases have been stagnating at around 1.5–2% per year, which can hardly compensate for the rapid growth in demand. Due to this structural asymmetry, the relative climate impact of aviation will increase significantly in the future even if absolute emissions remain constant.[12]

These factors make aviation a particularly complex field for sustainable transformation. At the same time, it also offers great potential for systemic change due to its innovative strength and global relevance.

Historical Development of the Sustainability Debate in Aviation

In the early decades of civil aviation, the focus was primarily on technological development, safety and economic growth. Following the end of World War II, civil aviation experienced unprecedented growth. Wartime innovations in aeronautics were repurposed for commercial use, leading to the proliferation of long-range aircraft and the establishment of international air routes. The formation of the ICAO in 1944, under the Chicago Convention, provided a regulatory framework that emphasised safety, efficiency and international cooperation. These principles would later intersect with sustainability goals.

The introduction of jet engines in the 1950s, particularly with the launch of the de Havilland Comet and later the Boeing 707, revolutionised air travel. Mass, fast and comfortable transport across continents was possible for the first time. Jet aircraft offered more reliable service but also introduced new environmental challenges. Jet engines consumed more fuel and emitted higher levels of emissions compared to their piston-engine predecessors. Environmental aspects hardly played a role during this phase, neither in public perception nor in the strategies of companies in the aviation sector.

Although the aviation industry was largely focused on growth and modernisation, the 1960s witnessed the first stirrings of environmental consciousness. The publication of Rachel Carson's *Silent Spring* in 1962, while not directly related to aviation, catalysed a broader environmental movement that would eventually influence all sectors, including air transport.[13] Concerns about noise pollution began to surface, particularly in communities near major airports. The increasing frequency and volume of jet traffic led to public complaints and the first calls for noise abatement measures.[14] In response, some governments and airport authorities began to implement noise regulations. For example, London Heathrow introduced night-time curfews and operational restrictions.[15] These early policies were not framed in terms of sustainability but reflected a growing recognition of aviation's social and environmental footprint. During this period, aircraft manufacturers and airlines also pursued improvements in fuel efficiency not for environmental reasons, but primarily to reduce operational costs. The development of more aerodynamic airframes, lighter materials and more efficient engines contributed to incremental gains in energy use per passenger- or per tonne-kilometre.[16]

With the strengthening of the environmental movement in the 1970s, triggered by reports such as *The Limits to Growth* by the Club of Rome,[17] the ecological impact of aviation became an even greater focus of attention. However, aircraft noise continued to be a central issue. The ICAO responded with the introduction of noise limits, which defined binding international standards for aircraft noise for the first time (see Chapter 5).

At the same time, the 1973 oil crisis led to an economically motivated rethinking with regard to fuel consumption. Manufacturers such as Boeing and Airbus began to develop further aerodynamically optimised aircraft and to cooperate with engine manufacturers to build more efficient turbines.[18] Routings in the flight schedules were optimised, the number of connections reduced and larger aircraft with a higher seating density were deployed. However, the motivation was primarily economic in nature and ecological aspects continued to play only a subordinate role. With the Kyoto Protocol of 1997, aviation was included in the global climate discourse for the first time. Although international aviation remained exempt from national emissions targets, the need for international solutions was recognised. Since then, ICAO has taken on a central role in the development of global environmental standards (see Chapter 5), for example through the CO_2 emissions standard for aircraft and the market-based system CORSIA (Carbon Offsetting and Reduction Scheme for International Aviation).

Another milestone was the Paris Climate Agreement of 2015. In the final document, the parties to the United Nations Framework Convention on Climate Change (UNFCCC) agreed to reduce greenhouse gas emissions in order to limit the global temperature increase to well below 2°C and set themselves the goal of keeping it below 1.5°C above pre-industrial levels.[19] In addition the Parties agreed to offset anthropogenic greenhouse gas emissions, i.e., to strive for climate neutrality in the second half of this century.[20] As emissions from aviation are anthropogenic, they are covered by the goals of the Paris Agreement, even if they are not explicitly mentioned.

With the adoption of the United Nations' Sustainable Development Goals (SDGs) in 2015, sustainability became an integral part of international policy and corporate governance. This applies to the aviation industry, for example, in the areas of climate protection (as SDG 13), infrastructure (SDG 9) and responsible consumption (SDG 12).

Current Status of Sustainability in Aviation

The transformation of aviation towards a more sustainable future involves a large number of stakeholders, some of whom have diverging interests. A differentiated view of the stakeholders is therefore essential in order to obtain an overview of the current status of sustainability activities, understand conflicting objectives and develop suitable measures.

In recent years, the public perception of aviation has changed significantly with regard to its environmental impact. The Swedish movement Flygskam ("flight shame") in particular has attracted international attention and raised awareness of the climate impact of air travel.[21] This has led to an increasing societal discourse on environmentally friendly mobility behaviour. Despite these developments, private traveller behaviour remains ambivalent. Many customers are interested in carbon offsetting and prefer "sustainable" airlines, for example by paying attention to environmental labels (e.g., Atmosfair Airline Index 2024).[22] However, actual booking decisions often do not reflect these preferences.[23] This phenomenon is described as the attitude-behaviour gap. Consumers' environmental attitudes are not reflected in corresponding behaviour. Although many passengers describe sustainability as an important criterion, they ultimately make their booking decisions primarily based on price and comfort. The gap between ecological attitudes and actual buying behaviour is caused by determinants such as values, beliefs, attitudes and social influences. Purchasing decisions are not based solely on ecological convictions, but also on factors such as self-interest, short-term thinking, social imitation and scepticism about the credibility of green offerings. More sustainable options, such as choosing lower-emission connections, are often associated with higher costs or less convenience, which makes practical implementation more difficult.[24]

At this point, it should be mentioned that survey results on sustainability-related topics are often subject to the influence of socially desirable response behaviour, the so-called social desirability bias. This effect describes the tendency of respondents to make statements that appear socially accepted or morally desirable, even if they do not correspond to actual behaviour. Particularly in the area of environmental and climate protection, people tend to overemphasise their environmental responsibility. This phenomenon makes it difficult to assess individual willingness to act in a valid way.[25]

Stakeholders in Aviation and Their Sustainability Requirements

More and more firms are systematically integrating emissions, e.g., from business travel, into their carbon footprint balance. The disclosure of emissions in corporate balance sheets is currently not mandatory worldwide but is already standard or is becoming increasingly common in certain regions and jurisdictions. This is particularly the case where there are strict ESG and climate reporting obligations (see Chapter 7). While Europe is leading the way, reporting in the UK and USA is still voluntary. The UK is planning to follow the EU Corporate Sustainability Reporting Directive (CSRD) with the UK Sustainability Disclosure Standards (SDS).[26] The reported emissions are not only generated directly within the company, but along the value chain, e.g., through business-related air travel. Increasing regulation and growing demands from investors and customers are prompting companies to implement new control mechanisms, e.g., the CSRD in the EU. The CSRD obliges companies to prepare standardised sustainability reports. The aim is to present environmental, social and governance aspects just as transparently and bindingly as key financial figures. Companies must disclose how their business activities affect the environment and society and vice versa. As a result, individual CO_2 budgets are being allocated to teams and employees and the systematic avoidance of travel, e.g., through video conferencing, is being recommended. Some companies even have offsetting obligations for the remaining air travel. These developments have far-reaching consequences for the aviation industry. The business travel sector, which historically generated high margins, e.g., through high-priced tickets that can be flexibly rebooked and purchased at short notice, is increasingly being substituted or called into question. Airlines must therefore realign their business model by focusing more on leisure travellers or more dynamic pricing models. The structural change in demand as a result of the decarbonisation goals of many large companies poses considerable challenges for the industry in the long term.

Many investors are intensifying pressure on companies to establish sustainable business models. Institutional investors in particular are now systematically integrating ESG criteria into their investment decisions. This has far-reaching consequences for the aviation industry. Access to capital is increasingly linked to the sustainability performance of airlines. Companies that do not communicate their climate strategies credibly or do not report emissions transparently risk reputational damage and the withdrawal of capital by ESG-oriented funds. At the same time, green bonds and sustainability-linked loans are gaining in importance. These financing instruments offer airlines the opportunity to acquire capital specifically for low-emission technologies albeit on condition that specific sustainability targets are met.[27] Reliable and standardised disclosure of climate-related risks in accordance with frameworks such as the CSRD is now a prerequisite for access to the capital market. Failures in ESG performance can not only increase financing costs, but also lead to exclusions from important indices or funds. The aviation industry is therefore under growing demand to further develop its business models in a climate-compatible manner.

Employees increasingly expect aviation companies to demonstrate a credible commitment to environmental issues. Particularly in light of climate change, a large number of employees are calling for tangible measures to reduce CO_2 emissions and use sustainable technologies, for example by switching to fuel-efficient aircraft. Internal processes such as the approval of business trips and the avoidance of unnecessary travel are also scrutinised. For many employees, environmental responsibility is part of corporate ethics as they expect transparency about climate targets, participation in sustainability initiatives and training programmes on ecological topics. A credible environmental policy not only increases identification with the employer, but also boosts motivation and reputation.

Airlines are at the centre of the global sustainability debate, as they are both significant climate-forcing emitters (CO_2 and non-CO_2) and at the same time potential enablers for climate-friendly mobility. According to the IATA, around 350 airlines are now committed to the goal of net zero emissions by

2050 in line with the Paris Agreement.[28] To achieve this goal, many airlines are investing in fleet modernisation and non-fossil fuels. At the same time, airlines are under considerable economic pressure. The market is characterised by volatile margins that are heavily dependent on geopolitical and economic developments. This makes long-term investment decisions difficult. In addition, there is a high capital requirement to finance technological change, e.g., new aircraft. Another issue is the lack of a level playing field. While European airlines are increasingly subject to stricter emissions policies such as the EU emissions trading scheme (ETS), competitors from other regions enjoy much more lax framework conditions. This distortion of competition makes the global decarbonisation of aviation more difficult and could lead to carbon leakage.[29]

Airports contribute significantly to decarbonisation as infrastructure providers. Many airport operators are currently investing in the expansion of infrastructure for SAF, including special refuelling tanks, pipeline connections and logistics systems. At the same time, the ground fleet is gradually being electrified, for example with electric tugs, buses or baggage carts, in order to reduce local emissions on the apron (see also Chapter 6). Airports are also increasingly relying on modern technologies in the area of energy efficiency. LED lighting, smart building technology and the expansion of photovoltaic systems are helping to minimise electricity consumption and CO_2 emissions from facilities. At the same time, airports are under pressure from society. Air and noise pollution, competition for land use and biodiversity aspects regularly lead to conflicts with local residents and environmental organisations. Social acceptance of new development plans is therefore increasingly dependent on credible sustainability strategies and participatory planning processes.

Ground Handling Service Providers are helping to reduce the carbon footprint of aviation and make the aviation industry more climate friendly. Activities such as baggage handling, passenger transport and aircraft refuelling are increasingly being performed in an environmentally friendly manner through the use of electric or hydrogen-powered ground vehicles, efficient handling processes and sustainable energy supplies. Innovative taxiing methods such as taxi bots or single-engine taxiing can reduce emissions and improve air quality while lowering noise levels in the airport environment. The decarbonisation of air traffic requires cross-sector cooperation. As part of a cooperative approach, ground handling companies see themselves not only as technical service providers, but also as strategic partners for a climate-neutral future.

Air Navigation Service Providers (ANSPs) such as DFS (Germany), NATS (UK) and the FAA (USA) are essential stakeholders in the debate on sustainable aviation. Their primary goal is to ensure safe, efficient and orderly air traffic. At the same time, they are facing increasing pressure to make their contribution to reducing negative climate impacts, for example through optimised flight routes, shorter holding patterns and the introduction of free route airspace. However, these environmental goals are sometimes in conflict with traditional safety and capacity requirements. For example, environmentally favourable direct flights or flexible routing can be more complex to coordinate and more difficult to reconcile with existing airspace management.

Non-Governmental Organisations such as Transport & Environment (T&E) and Greenpeace are among the most influential civil society players in the debate on the ecological transformation of aviation. Through campaigns, studies and political lobbying, they consistently call for stricter regulations. These include the introduction of a kerosene tax, bans on short-haul flights and ambitious emission limits (see Chapter 5). At the same time, they are committed to promoting climate-friendly alternatives such as rail transport.[30] Another focus of their work is criticising carbon offset programmes, which they often consider to be non-transparent and ineffective. Organisations like Greenpeace call for strict standards to prevent Greenwashing and emphasise that genuine emission reductions must take precedence over offsetting mechanisms.[31] Through public campaigns such as the claim for reduction of business and short-haul-flights or the capping of airport expansion, environmental organisations are increasingly influencing public opinion and thus also political decision-making at national and supranational level.

One of the industry's key problems is the disparity in regulation at a global level. The European Union in particular is going beyond ICAO standards with the EU emissions trading scheme (EU ETS), the ReFuelEU aviation package and binding SAF quotas for airlines by 2050 as an ambitious framework. However, these regulatory measures are not globally harmonised and therefore have the following consequences:

- Distortion of competition: Strict environmental regulations in Europe can lead to disadvantages for airlines compared to competitors in less regulated markets like the Middle East.
- Carbon leakage: Shifting emissions to less regulated markets rather than actually reducing them, sometimes even leading to a higher overall level of emissions.
- No global scalability: Fragmented international regulatory approaches such as different SAF quotas hinder globally scalable business models. At the same time, long-term planning security is needed, i.e., stable framework conditions for investments in new technologies.

International harmonisation of regulatory approaches, for example via ICAO or multilateral agreements, is therefore essential for an effective transformation of the industry. At the same time, such coordination processes are inherently complex and time-consuming, requiring sustained political commitment over the long term.

Sustainability beyond Climate Protection

Sustainability in aviation is often reduced to ecological aspects, especially CO_2 emissions. Due to increasing public criticism of frequent flying, low-priced tickets and Greenwashing, there is growing pressure to take a more comprehensive approach to sustainability. A holistic understanding of sustainability includes justice, economic and social participation as well as ethical responsibility. These dimensions are crucial to ensuring the long-term social acceptance of aviation. Important aspects are:

- Social participation: In many regions and particularly in the Global South aviation is a vital access point to healthcare, education and economic development.
- Affordability: In recent decades, the accessibility of air travel has improved considerably worldwide. Key drivers have been the liberalisation of air transport markets, particularly in Europe and North America, as well as the rise of low-cost carriers, which appealed to broader sections of the population with lower prices and simplified booking models. At the same time, technological innovations, digital distribution channels and the expansion of regional airports led to increased availability of flight connections. Socioeconomic factors such as global economic growth and the rise of a travelling middle class in emerging countries also contributed to the growing demand. At the same time, environmental costs are currently being externalised and there is a risk that social standards will be undermined.
- Accessibility in general: Airports and airlines must support people with reduced mobility e.g., through wheelchair-accessible infrastructure, boarding assistance and trained staff.

Sustainability also includes responsibility towards employees. The aviation industry faces several challenges in this regard:

- Working conditions in the aviation industry (among pilots, cabin crew, ground personnel and air traffic controllers) have significant implications for occupational health, operational safety and workforce sustainability. Numerous studies indicate high levels of fatigue, psychological strain and sleep disruption, especially among cockpit and cabin crews.[32] These factors are associated with reduced cognitive performance, impaired decision-making and increased risk of errors, potentially jeopardising both employee wellbeing and flight safety. Moreover, the prevalence of irregular shift schedules, night work and transmeridian travel contributes to circadian misalignment and long-term health risks. For example, understaffing and overtime pressure exacerbate physical and mental fatigue, further increasing the likelihood of operational incidents.[33]

- In Europe, there are a large number of trade unions in civil aviation, with employees in each country often being represented by several unions. The tightening global supply of e.g., pilots, strengthens the bargaining position, which has led to an enhanced negotiating power for sector unions. The fragmented nature of labour laws in Europe makes negotiations more complicated for companies operating in multiple countries. However, unions are challenged by more globalised airline business models, which seek to minimise labour costs by deploying staff in markets with lower costs and less powerful unions.
- Despite increasing social awareness, women and minorities are still significantly underrepresented in the aviation industry, particularly in technical professions (e.g., maintenance, flight management) and in management positions. The proportion of women among pilots worldwide is around 4%, and in management positions at large aviation companies it is often in the single-digit percentage range.[34] This structural underrepresentation not only leads to a gender equality deficit but also constitutes a missed opportunity for innovation and decision-making potential. Studies show that diverse teams in technology and management contribute to better problem-solving, greater customer proximity and economic resilience.[35] Against this backdrop, programmes to promote gender equality, cultural diversity and an inclusive corporate culture have become important. Initiatives such as IATA's "25by2025" or diversity charter models at national level aim to systematically change recruitment, training and promotion practices.[36] In the long term, such measures contribute to the sustainable transformation of the industry, both socially and economically.

Conflicting Goals and Necessary Transformation in Aviation

The aviation industry is facing a twofold transformation: on the one hand, it must drastically reduce its ecological impact and, on the other, ensure its economic viability and social acceptance. Sustainability encompasses more than just climate protection. Social justice, accessibility, fair working conditions and ethical responsibility are integral components.

However, the sustainable transformation of aviation is characterised by conflicting interests that touch on technological, economic, social and political dimensions. Modern, fuel-efficient aircraft and technologies such as SAF are seen as key levers for decarbonisation. As SAF production costs are currently many times higher than those of fossil fuel alternatives the lack of economic viability inhibits investment and leads to an innovation dilemma: without scaling, prices will not fall, and without a price reduction, there will be no scaling.[37]

Aviation constitutes a highly interconnected global sector whose emissions have a cross-border impact. Unilateral national approaches to regulation, e.g., through SAF quotas, lead to distortions of competition and carbon leakage. Like in many other industries the fragmentation of regulatory frameworks undermines the effectiveness of global climate protection measures. A decoupling of growth and emissions has not yet been achieved, which is already leading to discussions about a growth limit in countries such as the Netherlands.[38] Measures such as the internalisation of environmental costs make ecological sense, but can have a socially regressive effect. Low-income households are disproportionately burdened. A socially acceptable environmental policy therefore requires targeted mechanisms.

The transformation of aviation is not a guaranteed success. It is not a linear process, but a complex balance of competing interests. An integrative strategy must not only recognise conflicting targets, but also actively manage them through technological innovation, international cooperation, political control and social compensation measures. It requires courage, innovative strength and international cooperation. Only in this way can aviation make its contribution to achieving climate targets without losing social acceptance. The coming years will be decisive in setting the course for climate-friendly, socially just and economically viable aviation.

Key Takeaways

- Aviation faces a fundamental sustainability dilemma as it must reconcile continued growth in global mobility with the urgent need to reduce its environmental and social impacts. This tension makes sustainability a core strategic challenge rather than a peripheral issue.
- The climate impact of aviation extends beyond CO_2 emissions, as non-CO_2 effects caused by contrails and nitrogen oxides significantly amplify its overall contribution to global warming. Effective sustainability strategies must therefore address both direct and indirect climate effects.
- Long investment cycles and high capital intensity slow down transformation, as aircraft and infrastructure decisions lock in technologies for several decades. This structural inertia limits the speed at which low-emission innovations can be deployed at scale.
- Fragmented international regulation creates risks of competitive distortion and carbon leakage, particularly between regions with strict climate policies and those with more lenient frameworks. Global coordination through institutions such as ICAO is essential for effective and fair decarbonisation.
- Sustainability in aviation is multidimensional, encompassing not only environmental protection but also social equity, accessibility, labour conditions and diversity. Ignoring these social dimensions risks undermining public acceptance and long-term industry resilience.
- Technological innovation alone is insufficient to achieve climate-neutral aviation, as efficiency gains are partly offset by growing demand and rebound effects. A successful transformation requires integrated approaches combining technology, regulation, market mechanisms and international cooperation.

Notes

1 Schmithüsen (2013).
2 Brundtland (1987).
3 Elkington (1998).
4 UN (2004).
5 UN (2015a).
6 E.g., Fraport (2023).
7 IATA (2025b).
8 IATA (2025a) and Oliver Wyman (2025).
9 It was not until the third stage of liberalisation of the EU internal aviation market in April 1997 that cabotage was completely deregulated. Since then, all EU airlines have been allowed to operate domestic flights (cabotage flights) within another EU Member State.
10 See Jevons (1866), Smil (2017) and Alcott (2005). The Jevons paradox, which was first stated by the economist William Stanley Jevons in *The Coal Question* in 1866, argues that improvements in the efficiency of resource utilisation can lead paradoxically to an overall increase in the consumption of this resource. Empirical studies support this phenomenon. For example, Smil (2017) notes that the specific fuel consumption of jet engines has halved since the 1960s, but that global aviation emissions continue to rise due to the increased number of flights.
11 ACI (2025) and Allianz (2025).
12 ICCT (2022) and ICAO (2024).
13 Carson (1962).
14 Bednarek (2016).
15 Heathrow Airport (2025).
16 ICCT (2009).
17 Meadows et al. (1972).
18 ICAO (2009).
19 UN (2015b).
20 Anthropogenic refers to processes or materials that are originating from human activity, especially those that influence the environment or climate, such as pollution or greenhouse gas emissions.
21 Gössling et al. (2019).
22 Atmosfair (2024).
23 Patino-Rataza and Suau-Sanchez (2025).
24 Higham et al. (2016), Groening, Sarkis and Zhu (2018), Colantuono et al. (2025) and Kormos and Gifford (2014).
25 Eagly and Chaiken (1993).
26 In Japan, companies above a certain size report emissions, particularly in the financial and industrial sectors. There are also initial reporting obligations for large companies in South Korea.
27 ICMA (2023).
28 Around 350 airlines worldwide have made a commitment to the goal of net zero emissions by 2050 as part of the IATA initiative "Fly Net Zero" at the IATA Annual General Meeting, June 2025. Originally, almost 300 members signed the resolution at the IATA General Assembly in Boston in 2021.
29 Wei and Kallbekken (2024).
30 T&E (2025).
31 Greenpeace (2022).
32 Çevik, Elfvengren and Dahanayake (2022) and EASA (2024).
33 See ICAO (2025) and Wen et al. (2023). From a policy perspective, these findings underscore the necessity of robust Fatigue Risk Management Systems (FRMS), evidence-based duty time regulations and comprehensive mental health support

frameworks. In the context of growing demand for air travel and global pilot shortages, addressing these challenges is essential for ensuring a safe, resilient and socially sustainable aviation ecosystem.

34 ICAO (2023).
35 OECD (2024).
36 IATA (2025c).
37 Policymakers try to address this problem by the use of different instruments, like mandates, see Chapter 5.
38 Faber et al. (2022).

References

ACI – Airport Council International (2025) Joint ACI World-ICAO Passenger traffic Report, Trends, and Outlook; https://aci.aero/2025/01/28/joint-aci-world-icao-passenger-traffic-report-trends-and-outlook/

Alcott, B. (2005) Jevons' paradox; in: *Ecological Economics, Volume*54 (1); pp. 9–21; doi:10.1016/j.ecolecon.2005.03.020

Allianz (2025) High hopes, heavy footprint: Aviation's quest for climate-neutral skies; https://www.allianz.com/content/dam/onemarketing/azcom/Allianz_com/economic-research/publications/specials/en/2025/december/2025-12-02-aviation-AZ.pdf

Atmosfair (2024) Atmosfair Airline Index; https://www.atmosfair.de/wp-content/uploads/atmosfair-aai2024-ranking-background-final.pdf

Bednarek, J.R. (2016) *Airports, Cities, and the Jet Age: US Airports Since 1945*; Palgrave Macmillan, Basingstoke

Brundtland, G.H. (1987) Our Common Future: Report of the World Commission on Environment and Development; https://sustainabledevelopment.un.org/content/documents/5987our-common-future.pdf

Carson, R. (1962) *Silent Spring*; Houghton Mifflin Company, Boston

Çevik, I., Elfvengren, K., Dahanayake, A. (2022) Fatigue Risk Assessment of Cabin Crew for Aviation Transport Safety*; in*: Prentkovskis, O. et al (Eds.) *TRANSBALTICA XII: Transportation Science and Technology. Lecture Notes in Intelligent Transportation and Infrastructure*, pp. 765–774

Colantuono, R., Montanaro, A., Mazzanti, M., Mancinelli, S., Visintin, E.P. (2025) Flying toward sustainability? Investigating the attitude-behaviour gap among Italian air travellers; in: *Research in Transportation Business & Management*, Volume 62; 101466; doi:10.1016/j.rtbm.2025.101466

Eagly, A.H., Chaiken, S. (1993) *The Psychology of Attitudes*; Harcourt Brace Janovich, Fort Worth

EASA – European Union Aviation Safety Agency (2024) Study on the Analysis, Prevention and Management of Air Traffic Controller Fatigue; https://www.easa.europa.eu/sites/default/files/dfu/ATCO_Fatigue_Study_Final_report_D4C1_-_V4.1_final_21_May_2024.pdf

Elkington, J. (1998) Cannibals with Forks: The Triple Bottom Line of 21st Century Business; in: *Environmental Quality Management*, Autumn 1998; pp. 37–51; doi:10.1023/A:1006129603978

Fraport (2023) Fraport Annual Report 2022; https://www.annual-report.fraport.com/annual-report/2022/en/combined-management-report/combined-non-financial-statement/social-matters

Gössling, S., Hanna, P., Higham, J., Cohen, S., Hopkins, D. (2019) Can we fly less? Evaluating the 'necessity' of air travel; in: *Journal of Air Transport Management*, Volume 81; 101722; doi:10.1016/j.jairtraman.2019.101722

Grebe, S., Meijer, C., Király, J., Leestemaker, L. (2022) Impacts of a CO_2 ceiling for Dutch aviation; https://ce.nl/wp-content/uploads/2023/01/CE_Delft_220434_CO_2-ceiling_update_FINAL.pdf

Greenpeace (2022) Busted: 5 tricks the aviation industry plays on us to appear green; https://www.greenpeace.org/international/story/54040/busted-5-tricks-the-aviation-industry-plays-on-us-to-appear-green/

Groening, C., Sarkis, J., Zhu, Q. (2018) Green marketing consumer-level theory review: A compendium of applied theories and further research directions; in: *Journal of Cleaner Production*, Volume 172; pp. 1848–1866; doi:10.1016/j.jclepro.2017.12.002

Heathrow Airport (2025) Night Flights; https://www.heathrow.com/company/local-community/noise/operations/night-flights?utm

Higham, J., Cohen, S.A., Cavaliere, C.T., Reis, A., Finkler, W. (2016) Climate change, tourist air travel and radical emissions reduction; in: *Journal of Cleaner Production*, Volume 111; pp. 336–347; doi:10.1016/j.jclepro.2014.10.100

IATA – International Air Transport Association (2025a) Global Outlook for Air Transport – Trade, AI, and the energy transition; https://www.iata.org/en/publications/economics/reports/global-outlook-for-air-transport-december-2025/

IATA – International Air Transport Association (2025b) Chart of the Week 12 December 2025: Passenger and cargo grow despite supply constraints; https://www.iata.org/en/iata-repository/publications/economic-reports/passenger-and-cargo-grow-despite-supply-constraints/

IATA – International Air Transport Association (2025c) Advancing Diversity, Equity & Inclusion with 25by2025; https://www.iata.org/en/about/our-commitment/25-by-2025?utm

ICAO – International Civil Aviation Organization (2009) The Fuel Factor; https://www.icao.int/sites/default/files/sp-files/environmental-protection/Documents/Publications/6401_en.pdf

ICAO – International Civil Aviation Organization (2023) Aviation Personnel and Gender Statistics; https://www.icao.int/sites/default/files/APAC/Meetings/2023/2023-ADA2023/3-Documents/ICAO-aviation-personnel-and-gender_2023.pdf

ICAO – International Civil Aviation Organization (2024) Strategic Plan 2026–2050; https://www.icao.int/about-icao/Council/Pages/strategic-plan-2026-2050.aspx

ICAO – International Civil Aviation Organization (2025) Fatigue Management; https://www.icao.int/safety/fatigue-management

ICCT – International Council on Clean Transportation (2009) Efficiency Trends for New Commercial Jet Aircraft 1960 to 2008; https://theicct.org/wp-content/uploads/2021/06/ICCT_Aircraft_Efficiency_final.pdf

ICCT – International Council on Clean Transportation (2022) Vision 2050 – Aligning Aviation with the Paris Agreement; https://theicct.org/wp-content/uploads/2022/06/Aviation-2050-Report-A4-v6.pdf

ICMA – International Capital Market Association (2023) Sustainability-Linked Bond Principles – Voluntary Process Guidelines; https://www.icmagroup.org/assets/documents/Sustainable-finance/2023-updates/Sustainability-Linked-Bond-Principles-June-2023-220623.pdf

Jevons, W.S. (1866) *The Coal Question; An Inquiry concerning the Progress of the Nation, and the Probable Exhaustion of our Coal-mines*, 2nd edition; Macmillan and Co., London

Kormos, C., Gifford, R. (2014) The validity of self-report measures of proenvironmental behavior: A meta-analytic review; in: *Journal of Environmental Psychology*, Volume 40; pp. 359–371; doi:10.1016/j.jenvp.2014.09.003

Meadows, D.H., Meadows, D.L., Randers, J., Behrens III, W.W. (1972) *The Limits to Growth: A report for the Club of Rome's Project on the Predicament of Mankind*; Universe Books, New York

OECD – Organisation for Economic Co-operation and Development (2024) *Getting Skills Right: Bridging Talent Shortages in Tech: Skills-first Hiring, Micro-credentials and Inclusive Outreach*; https://www.oecd.org/en/publications/bridging-talent-shortages-in-tech_f35da44f-en.html

Oliver Wyman (2025) Global Fleet and MRO Market Forecast 2025–2035; https://www.oliverwyman.com/content/dam/oliver-wyman/v2/publications/2025/march/global-fleet-and-mro-market-forecast-2025-2035.pdf

Patino-Rataza, H., Suau-Sanchez, P. (2025) What drives passengers towards sustainable aviation? A segmentation study of travel behaviour and environmental concern in Spain; in: *Case Studies on Transport Policy*, Volume 21; 101539; doi:10.1016/j.cstp.2025.101539

Schmithüsen, F. (2013) Three hundred years of applied sustainability in forestry; in: *Unasylva*, Volume 64 (240); pp. 3–11; doi:10.3929/ethz-a-009955604

Smil, V. (2017) *Energy Transitions: Global and National Perspectives*, 2nd edition; Praeger, Santa Barbara, California

T&E – Transport & Environment (2025) Down to earth: Why European aviation needs to urgently address its growth problem; https://www.transportenvironment.org/articles/down-to-earth

UN – United Nations (2004) United Nations – The Global Compact & International Finance Corporation: Who Cares Wins – Connecting Financial Markets to a Changing World; https://documents.worldbank.org/pt/publication/documents-reports/documentdetail/280911488968799581

UN – United Nations (2015a) Transforming our world: the 2030 Agenda for Sustainable Development, Resolution A/RES/70/1; https://sdgs.un.org/2030agenda

UN – United Nations (2015b) Paris Agreement; https://unfccc.int/sites/default/files/english_paris_agreement.pdf

Wei, T., Kallbekken, S. (2024) Carbon leakage from aviation under the European Union Fit for 55 policies; in: *Transportation Research Part D: Transport and Environment*, Volume 132; 104269; doi:10.1016/j.trd.2024.104269

Wen, C.C.Y., Cherian, D., Schenker, M.T., Jordan, A.S. (2023) Fatigue and Sleep in Airline Cabin Crew: A Scoping Review; in: *International Journal of Environmental Research and Public Health*, Volume 20; 2652; doi:10.3390/ijerph20032652

3 Economic and Social Aspects of Aviation

Introduction

In the public debate, sustainability is often reduced to environmental aspects. However, the common definition of sustainability also includes economic and social aspects (see Chapter 2). Therefore, this chapter first outlines the economic relevance of aviation. Usually, economic effects are linked to the connectivity of a region. However, several definitions of connectivity exist, giving rise also to some methodological discussions. Moreover, different concepts are used for describing the economic impacts of aviation, ranging from economic impact analysis to computable general equilibrium models.

In the social dimension, the aviation industry as an employer in particular tries to increase the share of women in higher management positions as well as in professions like pilots and technicians. Moreover, Corporate Social Responsibility (CSR) also includes the support of local communities (Corporate Social Citizenship) as well as contributions to development and emergency response activities.

The Economic Dimension of Air Transport

Modern economies rely heavily on the division of labour to increase efficiency and wealth. International division of labour, in turn, is hardly possible without air transport, as it is the only option for time-critical shipments over longer distances. According to the International Air Transport Association (IATA), air cargo accounts for one third of world trade by value.[1] In addition, many passenger journeys – even with the growing importance of video-conferencing – are business related and serve purposes as varied as negotiations with (potential) customers, attendance at trade fairs or technical support and repairs.

If a region is well connected to other regions, the generalised costs of travel decrease,[2] thereby enhancing the competitiveness of firms located in this region and making the specific region more attractive for foreign investment. As an example, the European headquarters of large firms from South Korea (e.g., Samsung, Hyundai) are mostly located in regions that offer a

DOI: 10.4324/9781003608172-3

high number of flight frequencies to Seoul, e.g., Frankfurt (Germany) and London (UK).[3] However, although there is a high correlation between air transport connectivity and firms' locational decisions, causality works in both directions. In other words, when setting up branch offices and subsidiaries abroad firms often prefer regions that are well connected to their home country. At the same time, the existence of such entities also fosters the growth of air transport services between the respective cities. This is an example for the typical causality problem that researchers face when trying to empirically determine the significance of air transport for regional development. On the one hand, aviation promotes economic development; on the other hand, aviation markets benefit from regional growth.

How to Measure Connectivity?

Although the term connectivity is often used to describe an airport's significance for the regional economy, connectivity is difficult to define. First, many airports simply refer to the number of destinations which are served by direct flights. Although this can serve as an initial indication, frequency also plays a role, i.e., the number of connections per week and/or per day. Furthermore, the number of seats offered on these flights usually differs, which provides a further indicator of connectivity. Additional characteristics might be relevant, e.g., with respect to short- and long-haul flights or specific destinations. As a European example, London Heathrow offers the highest number of flights to North America, whereas Madrid-Barajas provides more connections to Latin America.[4] Moreover, from a business perspective, the economic relevance of the destinations is noteworthy. Direct flights to metropolitan areas like London, New York and Shanghai are particularly relevant to the economic development of a region, more so than flights to typical "sun and beach" destinations. Finally, indirect connections via hub airports are essential, especially for smaller airports. Here, too, it is not only the number of destinations and frequencies available with a transfer that counts, but also the total travel time. The latter is determined by the transfer time at the hub airport and by detour factors.[5]

Examples for Connectivity Metrics – IATA and ACI Europe

Several organisations calculate and publish connectivity metrics, e.g., the World Bank.[6] Moreover, in the economic literature specific metrics have been developed by researchers for dedicated purposes. In the following, connectivity indices published by two large industry associations, IATA and Airports Council International (ACI) Europe, are briefly introduced.[7]

IATA calculates a connectivity index with a worldwide coverage.[8] The index is based on the weekly number of seats on departing aircraft, multiplied by the relative size of the destination airport which is used as a proxy for its economic relevance as well as for transfer options.

ACI Europe publishes an "Airport Industry Connectivity Report" on a yearly basis. ACI Europe (2024a) shows direct connectivity (i.e., the number of direct flights) as well as indirect connectivity, using a specific method for incorporating additional travel time. In 2024, ACI Europe complemented its report with a "Power City Access Index" which also considers the economic importance of the respective destinations.

In 2024, according to ACI Europe, the five European airports with the highest direct connectivity were Istanbul Havalimani (IST), Amsterdam Schiphol (AMS), London Heathrow (LHR), Paris Charles de Gaulle (CDG) and Frankfurt (FRA). Based on the Power City Access Index, however, the sequence as well as the composition of the Top 5 European airports changes, with LHR at the top, followed by CDG, AMS, FRA and Madrid-Barajas (MAD).

For hub airports, the ACI Europe report also calculates a hub connectivity index, considering all potential connections via the respective airport. This index describes the competitiveness of the respective hub rather than its regional importance. However, there are some interdependencies between the hub connectivity and the direct connectivity of an airport.

The Role of Transportation Costs

Economic theory highlights the importance of low transportation costs in promoting trade.[9] In general, inflation adjusted air transport costs have fallen in recent decades, contributing to increased trade and thus economic efficiency. As an example, Figure 3.1 shows the development of inflation adjusted airfares for domestic round trips in the USA.

There are several reasons for the real cost reductions, including technological progress in aircraft design and operations as well as efficiency gains fostered by deregulation and privatisation of airlines. However, as some critics point out, the air transport industry also benefits from subsidies, tax exemptions and insufficient internalisation of external costs (see Chapter 5).

Aviation and Tourism

For many tourism destinations, the importance of air transport is apparent. Examples include island regions like the Canaries (Spain), Hawaii and the Seychelles, but also mainland destinations like Florida and California. Without convenient and affordable connections, the number of guest arrivals in these regions would drop dramatically. For example, in 2023 aviation's share of international arrivals from EU nationals was as high as 99% in

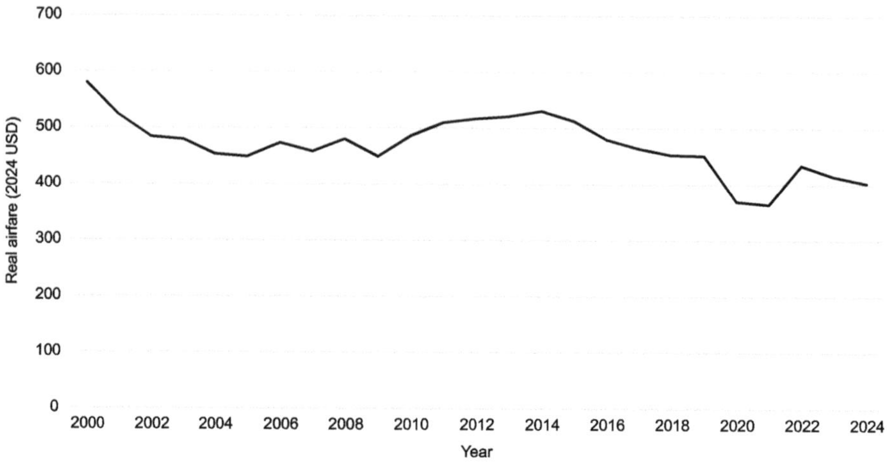

Figure 3.1 Real airfare for US domestic round trips. Figure created using data from Airlines for America (2025).

Malta, 94% in Cyprus and 85% in Greece.[10] However, there are also agglomerations like London, New York and Beijing which benefit from a large number of tourists that arrive by air.

Especially for islands and other rather remote regions, tourism is quite often by far the most important source of income. This might induce policy makers to support air transport links, in particular during off-peak seasons (see examples below) or to less popular destinations. From a sustainability perspective, however, there are also some detrimental effects, especially if the number of tourists is too large (overtourism). Potential concerns are environmental damage as well as social distortions, e.g., if tourism negatively affects the housing market for residents.[11]

Examples for Tourism-related Government Support

Malta, the smallest member state of the European Union, highly depends on tourism and most visitors arrive by plane. In its tourism strategy 2015–2020,[12] the Maltese government pointed out that Air Malta, as the back-then (state-owned and loss-making) national flag carrier,[13] was of pivotal importance for the tourism industry. In particular, unlike several foreign carriers it offered many connections also during the off-peak season, enabling a year-round operation of hotels and other tourism facilities.

In Israel, the government subsidises flights to the Red Sea airport Eilat during the winter, thereby trying to establish a year-round holiday destination.[14]

For a comprehensive analysis of the relationship between air transport and tourism it should also be considered that a better connectivity on the one hand will attract additional visitors from abroad, but on the other hand it might also increase the number of outbound tourists. This effect is stronger in bigger cities and agglomerations than in typical sun and beach destinations. Consequently, an analysis of regional economic effects (see paragraphs below on economic impact analysis) must not only focus on inbound tourism but should also look at the potential effects of additional outbound tourism.

The Passenger Perspective

From the perspective of private travellers, convenient and affordable air services facilitate many activities like visiting friends and relatives (VFR), going on vacation, participating in a student exchange programme or using medical services provided in specialised hospitals abroad. Especially for people living in remote regions, flight connections are crucial. Consequently, in many remote regions, not only is the airport infrastructure provided by the state but also services are subsidised by the respective jurisdictions.[15] Examples for such programs include Public Service Obligations (PSO) in the EU and UK, the Essential Air Service (EAS) programme in the USA and the Remote Air Services Subsidy (RASS) scheme in Australia.

The overall trend in airfare reductions and in particular the emergence of low-cost carriers, have led to a strong increase in the number of private travellers (and the number of trips per capita). Although this is sometimes referred to as the "democratization"[16] of air transport, critics point out that air transport can still be considered a privilege of wealthier people, be it on a global level (with a significantly lower share of air transport users in the Global South) or within one country. Moreover, Gössling and Humpe (2020) emphasise that a high share of air travel and especially of the related emissions is caused by a rather small number of frequent fliers.

Economic Impact Analysis

For many airports, an economic impact analysis has been commissioned by the airport operator or other entities, showing the regional (and sometimes national) significance of an airport in terms of employment, income/value added[17] and sometimes also tax revenue. Based on a largely standardised approach, these studies usually distinguish between direct, indirect, induced and catalytic effects. Figure 3.2 gives an overview, using employment effects as an example.

Direct effects measure the economic activity on the airport premises, i.e., the number of employees and/or their income. For example, more than 80,000 people work at the premises of Frankfurt Airport, being employed by approximately 450 firms (including the airport operator, airlines, aircraft maintenance providers, cargo warehouses, shops and restaurants in the terminal, etc.) and making the airport by far the largest place of employment in Germany.[18]

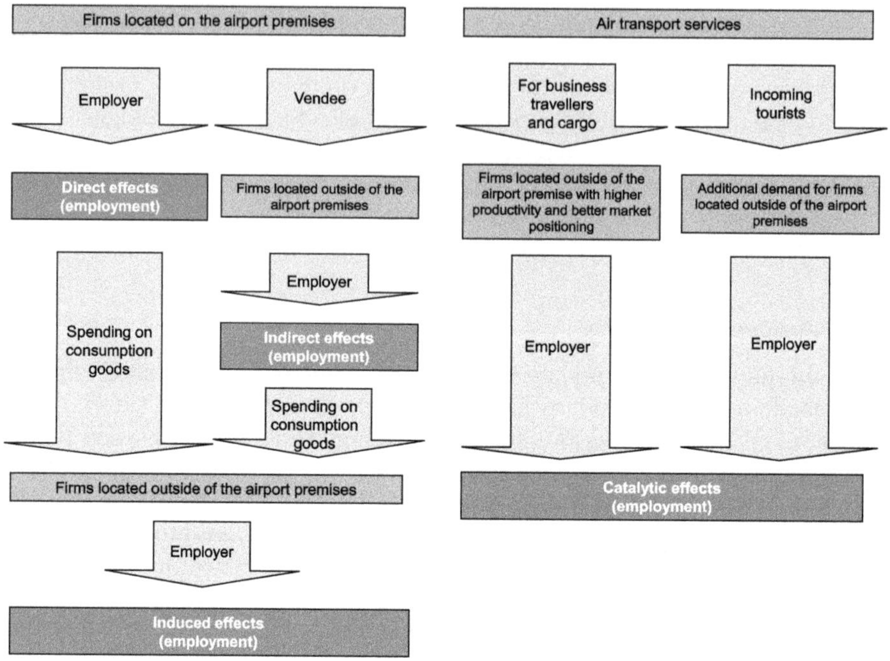

Figure 3.2 Overview of direct, indirect, induced and catalytic effects in economic impact analysis.

Indirect effects consider that firms operating at an airport purchase intermediate[19] as well as investment goods from firms located in the airport region. For example, the restaurant which is selling sandwiches in the terminal might purchase bread from a local bakery, and the airport operator might partially rely on local craftspeople to repair technical equipment in the terminal building. With the help of input-output tables additional backward links can be considered. To continue the example above, the ingredients for the bread produced by the local bakery might be purchased from local farmers – or they might be imported from other regions or countries – with only the first option additionally contributing to regional employment and income. These indirect effects might be calculated for regions of different size, reducing the comparability of studies prepared for individual airports. Moreover, some studies investigate the share of regional intermediate goods in a rather detailed way whereas other studies rely on approximations or average values.

Induced effects are based on the multiplier concept, which is a common element of "Keynesian" macroeconomic modelling.[20] Residents that directly or indirectly receive income generated by the airport operation will spend at least parts of this income in the airport region, thereby generating additional income and employment. Again, such an analysis might be rather detailed, considering for example the employees' places of residence,[21] or it might be based on aggregate numbers and assumptions.

Finally, catalytic effects are defined as all additional regional economic advantages enabled by the airport operation, including many effects that have already been mentioned in this chapter, e.g., tourism spending and a higher competitiveness of regional firms.

Selected Economic Impact Studies

Commissioned by ACI Europe (2024b), SEO Amsterdam Economics analysed the economic effects generated by ACI Europe's more than 500 member airports, using the general concept explained above. The results are summarised in Table 3.1. This study also provides results on the country level, with aviation having a higher (relative) share of economic activity in typical tourism destination countries like Malta and Cyprus.

ACI North America (2025) reflects a similar analysis for commercial airports in the USA and Canada. The study finds that the 487 commercial airports in the USA:

- Support 12.8 million jobs,
- Provide $619 billion in annual payroll,
- Produce $1.8 trillion of annual output.[22]

The ACI North America study distinguishes between on-airport activity, which basically comprises the direct effects explained above, the spending on incoming tourists in the airport region, which would be part of catalytic effects, and capital improvement projects, which can be considered as indirect effects generated by investment goods. Moreover, multiplier effects from all these activities are considered. Consequently, due to the different methodology and terminology, the results of this study on US airports cannot be compared directly with other, especially European studies.

Table 3.1. Economic impact of European airports in 2019 (ACI Europe 2024b).

	Employment (million jobs)	*GDP (billion 2022 Euro)*
Direct effects	1.843	121
Indirect effects	1.126	89
Induced effects	1.679	120
Catalytic effects (only tourism)	3.495	174
Total effects	8.143	504

Airport economic impact assessments are quite widespread, and the methodology is also used for other industries and activities, e.g., public transport providers and museums. However, there are also some conceptual peculiarities that have to be considered, especially when the respective numbers are used in public debates as justifications for airport expansion, airport subsidies and/or negative environmental effects.[23] First, to a large degree the respective studies describe the use of resources, but they do not discuss the necessity and/or efficiency of the respective activities or investment. For example, building an over-specified infrastructure would lead to higher income and employment in the airport region (e.g., construction workers and companies plus multiplier effects), and so would an over-staffing of the airport operator. Second, the economic impact assessment only looks at the airport itself and positive effects generated in other sectors. However, from an overall perspective there might also be some crowding-out effects. Assume, for example, that an airport region is characterised by full employment. Any additional airport employee would have to be poached from some other industry, leading to sector shifts rather than additional employment. There are specific economic models for incorporating such effects (in particular computable general equilibrium (CGE) models), but these models are usually rather highly aggregated.

Wider Economic Benefits

The term "wider economic benefits" of air transport activities is used quite often, but there is no uniform definition. Sometimes, the term is used more or less interchangeably with the term "catalytic effects". Other authors use the term wider economic benefits primarily for describing regional gains on the supply side of the economy, in particular cost reductions and an increasing productivity. One might argue, for example, that a high connectivity of a region facilitates the international exchange of knowledge and ideas, which increases productivity and the innovative capacity of firms located in the region.

Whereas – given available data – direct effects, tourism spending and indirect effects can be measured, cost reducing, market extending and productivity enhancing effects have to be estimated by sophisticated econometric models. For example, the study prepared for ACI Europe (2024b) shows that a 10% increase in connectivity of an airport leads to a 0.47% increase in regional GDP per capita and a 1.6% increase in employment. However, for the cargo market, no clear causality could be identified.

Additional Economic Effects along the Value Chain

The economic assessments presented above cover the activities of airports as well as connected activities, including tourist expenditures. However, this does not fully capture the aviation sector. In particular aircraft manufacturing and related activities might be added. According to ATAG (2024), the aviation industry in Europe supports 15 million jobs and contributes 1.2 trillion USD to European GDP. These numbers include aircraft and engine manufacturing, which accounts for 12% of direct employment in the sector.

Medical Flights, Firefighting and other Essential Activities

So far, this chapter has focused on the benefits of regular air transport, which is mostly provided on a commercial basis. Nevertheless, aviation might also help to overcome crisis situations, e.g., especially during the first months of the COVID-19 pandemic masks and other medical equipment were primarily transported by air.[24] Furthermore, in many cases transport by air is the only feasible option in emergencies. Examples include medical flights as well as firefighting missions and especially reactions by the international community in cases of national emergency, such as when hurricanes and tsunamis strike. Air transport is key to getting emergency supplies quickly to where they are needed. In order to enable these often live-saving flights, the appropriate infrastructure, as well as (stand-by) aircraft and personnel, have to be available.

The Social Dimension – Between Shareholders and Stakeholders

Principles of Corporate Social Responsibility

Adam Smith as well as other classical economists pointed out that firm owners basically are self-interested. Today, economic textbooks usually assume a profit-maximising behaviour of firms when analysing market conduct. However, in competitive markets firms have to be efficient, customer oriented and innovative in order to generate income for their owners/shareholders. Therefore, self-interested behaviour leads to a positive outcome also from a societal perspective. This sounds like a contradiction, but it is the result of a competitive market structure. Moreover, Adam Smith also emphasised the importance of morality for humans' decisions.

In management science, the moral "obligations" of firms are often described by using concepts like Corporate Social Responsibility.[25] There is no standard definition for this term, but one might distinguish between at least two aspects. First, as a minimum requirement, society would expect that firms obey the law, i.e., they do not engage in illegal activities like bribery, tax evasion and cartels, and they respect labour as well as environmental regulations. However, there might be grey areas, e.g., legal loopholes in tax laws that allow for a significant reduction of the tax bill, or different environmental standards between jurisdictions that enable a relocation of firms' activities to countries with less ambitious regulations. Second, firms might voluntarily go beyond the standards required by law. Examples include (additional) emissions reduction and improved working conditions. Moreover, a firm might support higher level objectives or societal goals by publicly promoting them and/or donating money or other resources for specific purposes. For these types of activities, often the term Corporate Social Citizenship is used. From a strategic management perspective, rule adherence is often considered to be part of a risk mitigation strategy, avoiding activities that might lead to negative consequences.

On the other hand, going beyond legal requirements can be part of a business opportunity strategy, i.e., an attempt to gain a competitive advantage. These strategy implications are further discussed in Chapter 8.

From an economic perspective, many of these activities might be in line with (long-term) profit maximisation. Concerning legal compliance, the economic analysis of law suggests that fines (or other penalties) should be set at a level that – given a specific detection probability – discourages such illegal activities, in other words, makes them unprofitable. Moreover, a sustainable behaviour of a firm might lead to competitive advantages if (1) customers value this behaviour and are willing to pay higher prices as a result, (2) the firm becomes more attractive for potential employees, (3) the firm gets better access to equity or outside capital and/or (4) a positive image of a firm or an industry creates a favourable environment in governments and parliaments. However, firm owners (or executives) might also sacrifice profits in order to become more environmentally sustainable and/or socially responsible. The motivation of the decision-makers could be purely internal or driven by societal expectations, thereby for example improving the personal image of the respective decision-maker.

Examples from the Aviation Industry

Along the aviation value chain, different options for dedicated CSR measures exist. As Chapter 9 deals specifically with sustainability marketing, this chapter focuses on labour aspects, procurement and Corporate Social Citizenship.

Concerning labour relations, one might argue that the share of skilled labour is rather high in aviation, leading to employment conditions above the average of the respective country. In certain professions, especially pilots and air traffic control officers, this is often fostered by the strong position of specialist unions. However, there are also many examples for labour conflicts, especially since the deregulation of the industry. In general, regulated (and state-owned) industries are often characterised by high wages and favourable employment conditions, facilitated by the absence of market competition. After the opening of aviation markets (especially US deregulation in 1978 and EU liberalisation in the 1990s), newcomers often benefitted from significantly lower labour costs, forcing incumbent airlines to also reduce costs in this area and causing many of the conflicts mentioned above.[26] Still, unions often criticise employment conditions at many low-cost airlines, including dedicated models for minimising social security contributions. Some airlines even took advantage of young pilots' need for gaining flight experience especially in situations of labour surplus, employing them at minimum wages or even letting them pay to fly an aircraft at a commercial service.[27]

In current human resource management, there is a general focus on equality and diversity, often giving a specific priority in higher management positions.[28] Again, this might also be in line with profit objectives as many studies suggest that diverse teams show a higher productivity and are more innovative than more uniform groups. In the aviation industry there are still several professions with a rather high share of male employees, in particular pilots,[29] but also technicians and engineers. Many firms aim at a more balanced workforce in these areas and also support initiatives with a broader outreach that try to attract more women to male-dominated professions, e.g., IATA's 25by2025 initiative.[30] Moreover, in line with the international nature of their business, airlines in particular are often characterised by a high share of employees from different foreign countries.

Corporate Social Responsibility also covers a firm's procurement decisions. Meanwhile, in many jurisdictions it is mandatory to address human rights and environmental protection along with the value chain, e.g., the EU's Directive 2024/1760 on corporate sustainability due diligence. Therefore, most firms aim at avoiding suppliers that violate human rights, in particular with respect to child labour and forced labour. Moreover, firms might procure goods that generate additional environmental or social benefits. Examples from the airline industry include goods with a fair-trade label or with a high share of recycled ingredients, e.g., for inflight catering.[31] However, it cannot be excluded that some airlines focus on products with a high visibility for consumers, giving rise to Greenwashing allegations (see Chapter 9).

Concerning Corporate Social Citizenship, airports often focus on the local and regional level, e.g., by supporting social and cultural activities.[32] Given that the negative environmental effects of airport activities primarily affect the local community, especially noise, and airports also to some degree depend on a favourable political framework at their location, these activities might also contribute to long-term economic objectives. Moreover, in many jurisdictions airports are fully or at least partially publicly owned, and the respective shareholders are therefore likely to be in favour of sponsoring activities that create regional benefits and support. On the other hand, airlines in particular are engaged also at an international level,[33] e.g., providing their specific resources for transporting medical supply to less developed countries.

From a customer perspective, airlines as well as airports serve a very diverse clientele, making the principle of non-discrimination a major element of their corporate culture. Cases of (alleged or actual) discrimination of certain customer groups usually lead to negative publicity for the respective airline. Moreover, in many jurisdictions specific legislation applies, including mandatory assistance for persons with reduced mobility (PRM), e.g., EU's Regulation 1107/2006.

Concerning public outreach, several airlines are supporting LGBTQ rights, e.g., by promoting Pride Month. Examples include media campaigns or even dedicated liveries, e.g., the "Lovehansa" aircraft of German airline Lufthansa.[34] Nevertheless, it should not be concealed that such activities might also evoke negative reactions in certain groups of the society.

Key Takeaways

- By facilitating trade, the air transport industry is essential for economic growth and prosperity, enhanced by a general long-term decline in real airfares.
- For private travellers air transport enables many valuable activities like visiting friends and relatives or recreation.
- Empirically, different connectivity metrics are used to describe the potential benefits to air transport users in a specific region.
- In many regions, tourism is the most important service industry and highly depends on air transport.
- Economic impact studies describe how many jobs are linked to the provision of air transport services and the magnitude of gross value added. However, additional benefits, e.g., a higher degree of innovation due to intensified international exchange, can only be estimated with the help of dedicated econometric models.
- Like in any other industry, airlines and airports are expected to adhere to the principles of Corporate Social Responsibility (CSR). Many respective measures are in line with the objective of long-term profit maximisation, but trade-offs are also possible.
- Specific CSR aspects in aviation include the promotion of a more gender-balanced workforce in professions like pilots and engineers as well as a commitment towards diversity in an international environment.

Notes

1 IATA (2025b).
2 Generalised cost of travel is a concept used in transport economics. It sums up the monetary costs of travel (e.g., the airfare) and the time-related costs, multiplying trip duration with the respective value of time.
3 For a comprehensive analysis see Bel and Fageda (2008).
4 ACI Europe (2024a).
5 The detour factor is defined as the ratio of the total flight distance of a connecting service and the respective direct flight.
6 Arvis and Shephard (2011).
7 A comparative discussion of the different approaches can be found in IATA (2020).
8 IATA (2020).
9 See for example Behar and Venables (2013).
10 Data source Eurostat (2025).
11 See for example Amrhein and Langer (2025).
12 Ministry for Tourism Malta (2015).
13 In 2024, loss-making Air Malta was replaced by a new state-owned airline in order to comply with EU rules on state aid. See Scicluna (2023).
14 Israel National News (2022).
15 An overview is provided in OECD (2011).
16 Calio (2018).
17 Value added is a term used in national income accounting. Subtracting the value of intermediate goods from a firm's revenue (without indirect taxes) yields the

gross value added, which consists of the income generated by the firm's activities (wages, rents, interest payments, dividends = net value added) and the depreciation on capital goods.

18 Fraport (2025).
19 An intermediate good is a good purchased by a firm which either becomes part of another product (e.g., wheat becomes part of a bread and bread becomes part of a sandwich) or is used up during the production process (e.g., energy that is used for baking a bread in a bakery or fuel that is burned while producing air transport services).
20 In macroeconomics, Keynesian models (named after British economist John Maynard Keynes) explain changes in output and employment primarily by changes in aggregate demand. As it is assumed that additional income leads to additional demand for consumption goods, the total effect in these models is larger than the initial change in demand (multiplier effect).
21 If an airport employee lives outside of the airport region, it can be assumed that his or her spending on consumption goods will to a large degree also be outside of the airport region.
22 "Output" basically is the same as revenue and therefore more than Gross Value Added which is used in European studies as an indicator.
23 For a critical discussion see for example Forsyth, Niemeier and Njoya (2021).
24 However, it has also been mentioned that air transport itself had a major role in the fast global spreading of the coronavirus from its origin in Wuhan, China throughout the globe. See for example Choi, Zou and Dresner (2022).
25 An overview is provided in Peeters et al. (2019).
26 For an early analysis see Alamdari and Morrell (1997).
27 Ricardo Energy & Environment (2019).
28 In some countries there are mandatory minimum shares for female board members in large firms.
29 See for example IATA (2024). Specific challenges for female pilots are discussed by Yanikoğlu et al. (2020).
30 IATA (2025a).
31 See for example Eurowings (2021).
32 See for example Heathrow Airport (2025).
33 See for example British Airways (2025).
34 Lufthansa (2022).

References

ACI Europe (2024a) Airport Industry Connectivity Report; https://www.aci-europe.org/downloads/publications/ACI%20EUROPE%20Airport%20Industry%20Connectivity%20Report%202024.pdf

ACI Europe (2024b) *The Economic and Social Impact of European Airports and Air Connectivity*; SEO, Amsterdam; https://www.aci-europe.org/downloads/resources/SEO%20Amsterdam%20Economics%20Study%20-%20The%20Economic%20and%20social%20impact%20of%20European%20Airports%20and%20air%20connectivity.pdf

ACI North America (2025) The Economic Impact of U.S. Commercial Service Airport in 2024; https://airportscouncil.org/wp-content/uploads/2025/03/ACI-NA-The-Economic-Impact-of-U.S.-Commercial-Service-Airports-in-2024.pdf

Airlines for America (2025) Data & Statistics; https://www.airlines.org/dataset/annual-round-trip-fares-and-fees-domestic/

Alamdari, F.E., Morrell, P. (1997) Airline labour cost reduction: post-liberalisation experience in the USA and Europe; in: *Journal of Air Transport Management*, Volume 3 (2); pp. 53–66; doi:10.1016/S0969-6997(97)00024-0

Amrhein, S., Langer, M. (2025) Resistance and power dynamics in tourist destinations: a study of Mallorca's anti-touristification protests through Bourdieu's theory of practice; in: *Journal of Sustainable Tourism*, Volume 34 (1); pp. 86–103; doi:10.1080/09669582.2025.2455397

Arvis, J.-F., Shephard, B. (2011) The Air Connectivity Index. Policy Research Working Paper 5722; The World Bank; https://documents1.worldbank.org/curated/en/859151468161649899/pdf/WPS5722.pdf

ATAG – Air Transport Action Group (2024) Aviation Benefits Beyond Borders; Geneva; https://aviationbenefits.org/media/e5ynn4x0/abbb2024_full_report.pdf

Behar, A., Venables, A.J. (2013) Transport costs and international trade; in: de Palma, A. et al. (Eds.) *A Handbook of Transport Economics*; Edward Elgar, Cheltenham and Northampton; pp. 97–115

Bel, G., Fageda, X., (2008) Getting there fast: globalization, intercontinental flights and location of headquarters; in: *Journal of Economic Geography*, Volume 8; pp. 471–495; doi:10.1093/jeg/lbn017

British Airways (2025) Charity and community investment; https://www.britishairways.com/content/en/de/information/about-ba/community-investment

Calio, N. (2018) The Democratization of Air Travel; https://www.airlines.org/nick-calio-the-democratization-of-air-travel/

Choi, Y., Zou, L., Dresner, M. (2022) The effects of air transport mobility and global connectivity on viral transmission: Lessons learned from Covid-19 and its variants; in: *Transport Policy*, Volume 127; pp. 22–30; doi:10.1016/j.tranpol.2022.08.009

Eurostat (2025) Tourism statistics – intra-EU tourism flows; https://ec.europa.eu/eurostat/statistics-explained/index.php?title=Tourism_statistics_-_intra-EU_tourism_flows

Eurowings (2021) New catering concept: Eurowings aims for more sustainability on board; https://newscloud.eurowings.com/en/new-catering-concept-eurowings-aims-for-more-sustainability-on-board/

Forsyth, P., Niemeier, H.-M., Njoya, E. (2021) Economic Evaluation of Investments in Airports: Recent Developments; in: *Journal of Benefit-Cost Analysis*, Volume 12 (1); pp. 85–121; doi:10.1017/bca.2020.31

Fraport (2025) Facts & Figures. The Airport and Its Workforce; https://www.fraport.com/en/our-group/about-us/facts—figures.html

Gössling, S., Humpe, A. (2020) The global scale, distribution and growth of aviation: Implications for climate change; in: *Global Environment Change*, Volume 65; 102194; doi:10.1016/j.gloenvcha.2020.102194

Heathrow Airport (2025) Funding and community partnerships; https://www.heathrow.com/company/local-community/funding-and-community-partnerships

IATA – International Air Transport Association (2020) Air Connectivity. Measuring the connections that drive economic growth; https://www.iata.org/economicsair-connectivity-measuring-the-connections-that-drive-economic-growth/

IATA – International Air Transport Association (2024) Gender diversity in aviation improving – but more to be done; https://www.iata.org/en/iata-repository/publications/economic-reports/gender-diversity-in-aviation-improving-but-more-to-be-done/

IATA – International Air Transport Association (2025a) Advancing Diversity, Equity & Inclusion with 25by2025; https://www.iata.org/en/about/our-commitment/25-by-2025/

IATA – International Air Transport Association (2025b) Value of Air Cargo; https://www.iata.org/en/programs/cargo/sustainability/benefits/

Israel National News (2022) Tourism Ministry offers airlines incentives to fly to Eilat; https://www.israelnationalnews.com/news/325706

Lufthansa (2022) Lufthansa takes off as "Lovehansa"; https://newsroom.lufthansagroup.com/en/lufthansa-takes-off-as-lovehansa/

Ministry for Tourism (Malta) (2015) National Tourism Policy 2015–2020; https://tourism.gov.mt/wp-content/uploads/2023/04/National-Tourism-POlicy-2015-2020.pdf

OECD – Organisation for Economic Co-operation and Development (2011) Air Transport Services in Remote Regions. *International Transport Forum*. Discussion Paper 13; Paris; https://www.oecd.org/content/dam/oecd/en/publications/reports/2011/05/air-transport-services-in-remote-regions_g17a1fe2/5kg9mq3xcrxx-en.pdf

Peeters, P., Bouwer, J., Bongaerts, R., Eijgelaar, E. (2019) Airlines and Corporate Responsibility: Issues and Challenges; in: Lund-Durlacher, D., et al. (Eds.) *Corporate Sustainability and Responsibility in Tourism. A Transformative Concept*; Springer, Cham; pp. 163–180

Ricardo Energy & Environment (2019) Study on employment and working conditions of aircrews in the EU internal aviation market; Commissioned by the European Commission, DGMOVE; Didcot; https://op.europa.eu/en/publication-detail/-/publication/97abb7bb-54f3-11e9-a8ed-01aa75ed71a1

Scicluna, C. (2023) Air Malta to close in March, new airline to replace it, PM says; Reuters; https://www.reuters.com/business/aerospace-defense/air-malta-close-march-new-airline-replace-it-pm-says-2023-10-02/

Yanikoğlu, Ö., Kiliç, S., Küçükönal, H. (2020) Gender in the cockpit: Challenges faced by female airline pilots; in: *Journal of Air Transport Management*, Volume 86; 101823; doi:10.1016/j.jairtraman.2020.101823

4 Environmental Aspects of Aviation

Introduction

The environmental impact of aviation is multifaceted, spanning from the immediate and locally harmful effects of aircraft noise and air pollution to the complex, long-term global challenge of climate change. Importantly, the aviation sector is not only a contributor to global warming, but also increasingly affected by it, for example through more frequent and severe turbulences, heat-induced limitations on aircraft performance and airport operations as well as increased weather-related delays and cancellations. In addition to these primary concerns, aviation also contributes to environmental problems through contamination of water and soil, the generation of large volumes of waste and land use, especially soil sealing. These diverse impacts differ in scale, complexity of their underlying mechanisms and regulatory response, but collectively define the environmental footprint of the aviation industry. This chapter provides a structured overview of the most significant environmental effects associated with air transport, highlighting their sources and consequences. Technological and operational mitigation approaches will then be discussed in Chapter 6.

Aviation Noise

Noise can generally be defined as unwanted and/or harmful sound. For residents living near airports, aircraft noise is often the most direct and personally intrusive environmental impact of aviation. It is an immediate sensory experience that can disrupt daily life (especially at night -time) and has been one of the primary drivers of public complaints, political debates and policy measures. A thorough understanding of the sources of aviation noise, its consequences for health and well-being and the complex framework for its measurement is therefore a key issue for the aviation industry.

Sources of Aircraft Noise

Aircraft noise originates from two primary sources: the propulsion system (engine noise) and the aircraft's airframe (aerodynamic noise). During high-thrust phases

DOI: 10.4324/9781003608172-4

such as take-off and initial climb, but also after landing when using reverse thrust, engine noise is the dominant factor for the total noise footprint. In contrast, during the low-power approach and landing, aerodynamic noise – generated by the turbulence from extended landing gear, flaps and slats – becomes more dominant. Because these critical noise-generating events are limited to the low-altitude Landing and Take-Off (LTO) cycle, their impact is usually a local phenomenon, and noise from the high-altitude cruise portion of a flight is not a significant factor.[1]

In addition to aircraft operations, ground activities at the airport also generate noise. This includes noise from various ground service vehicles, though their impact beyond the airport's perimeter is generally limited. But more significantly, engine tests conducted at maintenance facilities on airports can produce considerable noise, especially if they take place at night.

Consequences of Aviation Noise Exposure

To understand the impacts of aviation noise, it is essential to first distinguish between sound and noise. Sound is a physical phenomenon: an acoustic pressure wave that can be objectively measured using microphones. Noise, in contrast, can be best described as a subjective and psychological concept indicating sound that is perceived by an individual as unwanted, bothersome, or harmful, resulting in annoyance, a state of discomfort that can interfere with daily activities like conversation, concentration and relaxation.[2]

This psychological stress is compounded by direct physiological effects, particularly through sleep disturbance. Noise events at night can disrupt the natural sleep architecture, shifting the brain from deep, restorative sleep to lighter stages. Each of these arousals triggers a physiological response, leading to the release of stress hormones and an increase of blood pressure. Over the long term, chronic stress on the cardiovascular system is linked to an increased risk of developing diabetes, hypertension (high blood pressure) and ischaemic heart disease.[3] In this context, children are a particularly vulnerable group. Their chronic exposure to aircraft noise in educational settings can impair cognitive development, leading to reduced reading comprehension and memory deficits.[4]

Beyond health impacts, noise also has measurable economic consequences, primarily through the depreciation of property values.[5] Further economic effects include but are not limited to the costs of sound insulation programmes and potential negative impacts on local tourism. Finally, it is often neglected that noise also negatively affects wildlife.

Fundamentals of Noise Measurement

To regulate and manage a primarily subjective phenomenon like noise, the experience of "annoyance" must be translated into an objective and legally applicable framework. This is challenging because the human response to a sound event not only depends on measurable factors like volume and frequency but also on subjective factors like context and personal sensitivity. The fundamental unit for objectively measuring sound intensity is decibel (dB), which uses a logarithmic scale. The logarithmic scale is used because the human ear's sensitivity is not linear; it is highly sensitive to minor intensity differences at low sound levels, whereas at high sound levels, its ability to discern subtle variations decreases.[6] To further better align physical measurements with human perception, a frequency filter known as A-weighting is commonly applied, as the human ear is more sensitive to some frequencies than others. Measurements using this filter are then denoted as dBA or dB(A) and often simply referred to as "sound level".

Key Metrics for Aircraft Noise

Using dBA as measurement for sound intensity, a range of standardised metrics has been developed to quantify noise. These can be broadly categorised into two types: those that assess a single noise event (primarily used for aircraft certification, see Chapter 5) and those that measure cumulative noise exposure over time (typically used for assessing community impacts).[7]

The single-event metrics include:

- L_{max} (Maximum Sound Level): The highest dBA recorded during an aircraft overflight.
- SEL (Sound Exposure Level): A measure of the total acoustic energy of a single event, which accounts for both its intensity and its duration. The sound level is time integrated and normalised to a one-second reference period, capturing the cumulative sound energy over the interval during which noise exceeds a defined threshold (see Figure 4.1).[8]
- EPNL (Effective Perceived Noise Level): A metric used by the International Civil Aviation Organization (ICAO) for aircraft noise certification (also referred to as EPNdB). It builds upon the SEL but adds corrections for tonal qualities and their changes during a sound event.

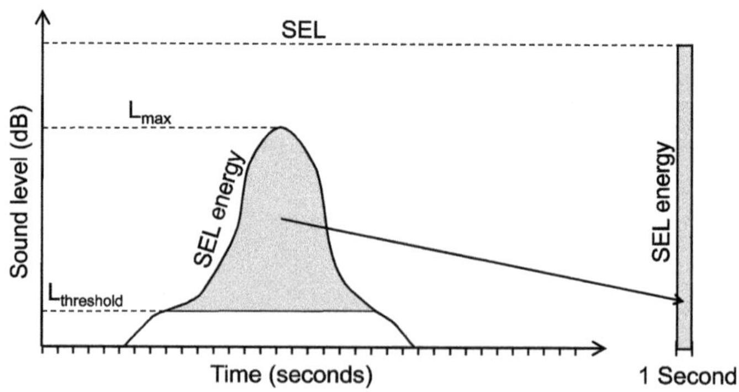

Figure 4.1 Maximum Sound Level (L_{max}) and Sound Exposure Level (SEL). Modified from FAA (2025b).

To assess the long-term impact on communities, cumulative exposure metrics are used, especially:

- L_{eq} (Equivalent Continuous Sound Level): Represents the cumulative effect of multiple noise events. It is a single numerical value representing the average of varying noise levels over a specified time interval.
- L_{dn} or DNL (Day-Night Average Sound Level): A 24-hour L_{eq} with a 10-decibel penalty applied to all noise events occurring during a defined night-time (see Figure 4.2).[9] It is the standard metric in the USA.
- L_{den} (Day-Evening-Night Level): The standard metric in Europe, similar to DNL but with a 5-decibel penalty for evening hours in addition to the 10-decibel night-time penalty.
- L_{night} (Night-Level): The L_{eq} for the night-time period only, making it a critical metric for assessing the risk of sleep disturbance.

These cumulative metrics are used by authorities to generate noise contours on maps (see Figure 4.3). These contours connect points of equal noise

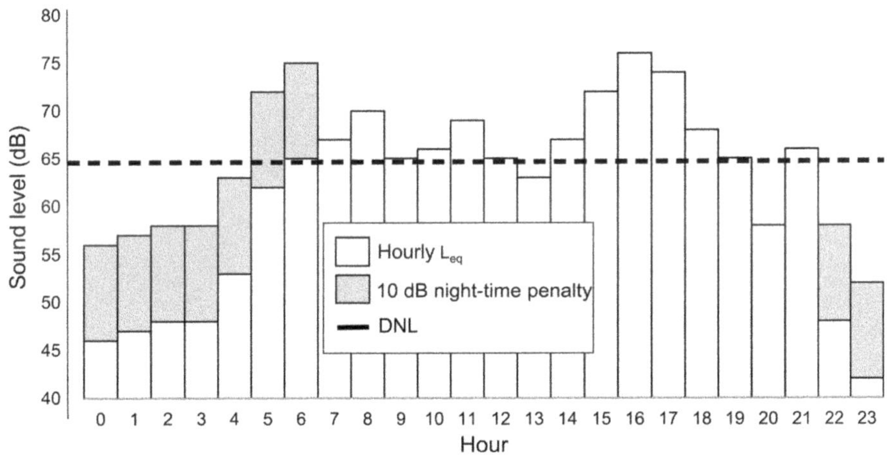

Figure 4.2 Day-Night Average Sound Level (DNL) based on hourly Equivalent Continuous Sound Levels (L_{eq}) and 10-decibel night penalties. Modified from FAA (2025b).

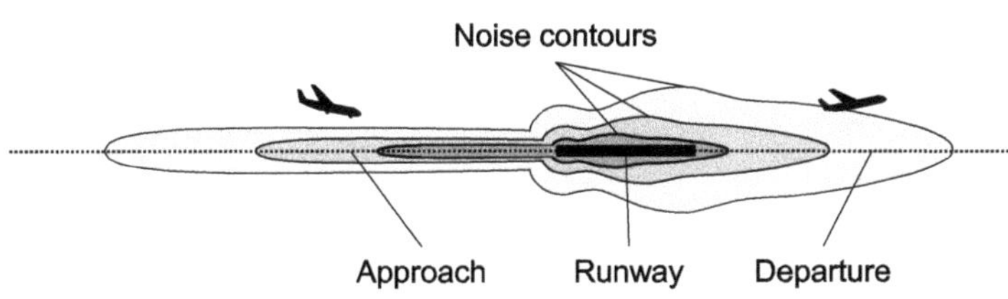

Figure 4.3. Schematic representation of noise contours. Based on FAA (2025b).

exposure (e.g., the L_{den} 55 dB contour) and are the primary tool for implementing land-use planning, sound insulation programmes and operating restrictions or noise abatement procedures around airports (see Chapter 5).[10]

Quantifying the Impact of Aviation Noise

While the annoyance level of a noise event and its aggregation is highly subjective, quantifying the number of people exposed to significant noise levels is essential for regulation and management. This quantification is, however, complicated by the use of different metrics and regulatory thresholds in different parts of the world and different scopes of the analyses. In Europe according to the European Aviation Safety Agency (EASA 2025a), the standard cumulative metric is the Day-Evening-Night Level (L_{den}), with a key threshold for adverse health effects at 55 dB (50 dB for L_{night}). The World Health Organization (WHO 2018) recommends even stricter thresholds with 45 dB L_{den} and 50 dB for L_{night}. In contrast, the USA use the DNL and typically report on the population exposed to a higher threshold of 65 dB. This makes direct international comparisons of exposure data challenging.

Noise Exposure for Different Regions

A comprehensive analysis of the population affected by aviation noise in Europe is provided by EASA (2025a).[11] In the pre-pandemic year 2019 EASA estimates that 3.8 million people were exposed to noise levels above L_{den} 55 dB and 1.5 million to night-time levels above L_{night} 50 dB. The most recent data for 2023 shows slightly lower numbers, with 3.4 million people exposed above L_{den} 55 dB and 1.45 million above L_{night} 50 dB, reflecting that traffic had not fully returned to 2019 levels (ca. 9% less) and that the noise energy per operation was further reduced (ca. 7% less).

Separately, EASA also reports on data submitted by member states under the EU's Environmental Noise Directive (END).[12] The currently most recent dataset reflects the traffic levels of the pandemic-affected year 2021, showing a lower exposure of 2.4 million people above L_{den} 55 dB and 0.7 million above L_{night} 50 dB. Based on this 2021 exposure data, the estimated annual health impacts include 649,000 highly annoyed people and 127,000 who are highly sleep disturbed.[13] The number of new disease cases attributable to this exposure is estimated at 500 annually for cardiovascular diseases and 200 for type 2 diabetes, leading to an estimated 600 premature deaths. Furthermore, more than 7,900 pupils may experience reading impairments.[14]

In the USA, using the higher threshold of DNL 65 dB, the number as well as the share of people considered to be significantly exposed is much lower than in Europe. For 2023, the Bureau of Transportation Statistics (BTS) reported approximately 397,000 people within this contour, representing just 0.12% of the population.[15] On a global level, ICAO (2022) estimated for the year 2018 for 319 major airports representing 80% of the global traffic an exposure of 37 million people to noise levels above 55 dB DNL.

Disproportionate Impact of Aviation Noise

A crucial aspect in assessing the impact of aviation noise is that it is consistently perceived as more annoying – and, thus, associated with more physiological as well as economic consequences – than noise from other transportation sources, such as road or rail traffic, at the same decibel level.[16] The higher annoyance is often attributed to the nature of aircraft noise, which consists of discrete, intense events that are less predictable and harder to avoid than the more continuous, ambient noise from ground transport. Underscoring this psychoacoustic difference, reports say that adverse health effects from aircraft noise are observed at exposure levels approximately 10 dB lower than for other sources. Consequently, despite affecting fewer people than ground transport modes like road and rail, the higher annoyance factor of aviation noise leads to a disproportionate overall impact.

The Outlook for Aviation Noise: Quieter Aircraft versus Growing Traffic

Technological progress in reducing aircraft noise, reflected also by increasingly stringent ICAO certification standards (see Chapter 5), has led to significant reduction in the noise energy produced by aircraft.[17] Consequently, modern jet aircraft are roughly 75% less noisy than the first models, and this downward trend continues with each new generation, whose noise footprint is estimated to be at least 15% smaller (and in the case of some new aircraft, up to 75% smaller) than its predecessor's.[18] For Europe, EASA (2025a) expects the average noise energy per flight to decrease by over 50% between 2019 and 2050.

However, these technological improvements at the aircraft level are offset by the effects of overall growth in air traffic. Despite each flight becoming less noisy, the projected increase in the total number of flights means that the overall noise exposure for the population around European airports is not expected to decrease significantly in the near future. The EASA (2025a) forecast mentioned above predicts that the number of people exposed to noise levels above L_{den} 55 dB will increase by 13% by 2030 before returning to slightly below 2019 levels by 2050. The outlook for night-time noise is similar, with a projected increase of 16% by 2030. This illustrates the central challenge of noise management, where per-aircraft efficiency gains are outweighed by overall industry growth.

This development might also change the management of the noise problem. Annoyance is often highest when communities are newly exposed to aircraft noise. While the construction of new airports or major runway expansions has become less frequent in developed regions like Europe, "new noise" can still be created through the redesign of flight paths or operational procedures. Overall, however, the challenge of managing the noise impact of large-scale airport capacity growth might shift more from established aviation markets to those in fast-growing, emerging economies.

Local Air Quality

Beside noise, the impact on local air quality represents the second major environmental challenge originating at the airport level and leading to considerable public health implications. While aviation's total emissions of air pollutants may be smaller than that of other transport sectors like road traffic across a whole metropolitan area, its emissions are highly concentrated within and around airport boundaries. This concentration creates local "hotspots" where ambient levels of harmful pollutants can be significantly elevated, possibly posing a direct health risk to both surrounding communities and the millions of personnel working at airports worldwide.[19]

Sources of Air Pollution from Aviation

The sources of air pollution in an airport environment are diverse, but they are primarily linked to the combustion of fossil fuels. While minor emissions can stem from fuel evaporation during handling, the primary source is fuel combustion in aircraft engines, their Auxiliary Power Units (APUs) and in the vehicles of the ground service equipment (GSE) fleet.

The Inherent Emissions from Hydrocarbon Combustion

Fundamentally, gaseous and particulate emissions (not only) from aviation result from the chemical reactions of combustion within the engines. In an ideal propulsion system, one would simply mix and ignite hydrogen (H_2) with oxygen (O_2) to release thermal energy, producing only water vapour (H_2O) as a byproduct. In practice, this idealised reaction is not achievable for two main reasons.

First, the energy is currently not supplied by pure hydrogen but is stored in hydrocarbons (jet fuel), which are primarily derived from fossil sources. The presence of carbon (C) in the fuel inevitably leads to the formation of carbon dioxide (CO_2) as one major combustion product. Second, neither the fuel nor the air used for combustion is pure. Aviation fuel contains various impurities, most notably sulphur, as well as a range of additives required for performance and safety. And the oxygen for the reaction is sourced from ambient air, which consists of approximately 78% nitrogen (N_2).

Thus, the combustion process in a jet engine is more complex. The combination of hydrocarbon fuel, fuel impurities and the high nitrogen content of the air – all subjected to high temperatures and pressures with varying combustion efficiency – inevitably results in a cocktail of undesirable emissions. Alongside carbon dioxide and water vapour, this includes nitrogen oxides (NO_x), sulphur oxides (SO_x), unburned hydrocarbons (HC), carbon monoxide (CO) and particulate matter (PM), which are the primary drivers of aviation's environmental impacts.[20]

In its 2019 Environmental Report, EASA (2019) cites a calculation by the Swiss Federal Office of Civil Aviation (FOCA) estimating the emissions from a typical twin-engine jet carrying 150 passengers on a one-hour flight. For such a flight, approximately 2,700 kg of fuel is burned, resulting in around 11,800 kg of emissions. Of these, 99.7% are accounted for by CO_2 and H_2O, with CO_2 contributing ~71.8% (~8,500 kg) and H_2O ~27.8% (~3,300 kg). The remaining 0.3% consists of NO_x, CO, HC and particulate matter or soot.

While the amounts of CO_2 and H_2O are directly proportional to the amount of fuel burned[21] – each kilogram of fuel produces about 3.15 kg of CO_2 and 1.24 kg of H_2O – the quantities of the other pollutants vary depending on engine efficiency and operating conditions, particularly the thrust setting.

Key Air Pollutants and Their Effects

While the main exhaust gas products, CO_2 and H_2O, are only relevant for the global climate, the other, less extensive emissions also have a direct impact on local air quality and human health. These primary pollutants can become secondary pollutants in the atmosphere through further chemical reactions.[22]

The key air pollutants and their impacts are as follows:[23]

- Nitrogen oxides (NO_x): This group of gases is formed under high-temperature and high-pressure conditions. NO_x is a potent respiratory irritant that can impair lung function. In the atmosphere, it is also a key precursor for the formation of secondary pollutants like ground-level ozone (the main component of smog) and secondary particulate matter.
- Particulate matter (PM): These are tiny solid or liquid particles, primarily emitted directly as soot from incomplete combustion. Of particular concern for human health are the very small particles like $PM_{2.5}$ and especially the ultrafine particles (UFP).[24] Due to their small size, they can penetrate deep into the lungs, enter the bloodstream and thus contribute to cardiovascular and respiratory diseases.
- Carbon monoxide (CO) and unburned hydrocarbons (HC): Both groups of pollutants are typical products of inefficient, incomplete combustion at low thrust settings, such as during ground taxiing. CO is a toxic gas that impairs the blood's ability to carry oxygen. HCs contribute to the characteristic smell at airports, and some of their compounds can be harmful to health. Both are also preliminary products for the formation of ground-level ozone.

- Sulphur oxides (SO_x): These emissions result from the combustion of sulphur naturally present in fossil-based fuel. SO_x can irritate the respiratory system and contributes to the formation of secondary particulate matter in the atmosphere.

ICAO Landing and Take-Off Cycle

To standardise the assessment of aircraft emissions near airports, ICAO (Annex 16, Vol. 2) has defined a reference Landing and Take-Off (LTO) cycle (see also Chapter 5). This cycle includes the aircraft's operational phases below an altitude of 3,000 feet (914 metres) and applies standard time in mode and thrust settings for engine certification and emissions modelling (see Table 4.1).

This reference LTO cycle is utilised by ICAO for certification processes, setting emission limits, technology comparisons and calculating local aircraft emissions at airports.[25] The emissions profile of an engine changes drastically during these phases.[26] During the high-thrust take-off and climb-out phases, the high temperatures and pressures in the engine core lead to the formation of large quantities of nitrogen oxides (NO_x). In contrast, during low-power ground operations like taxiing, the less efficient combustion process leads to higher emissions of unburned hydrocarbons (HC), carbon monoxide (CO) and particulate matter (PM).

Non-LTO Emissions at the Airport

While the LTO cycle standardises main engine emissions during flight-related phases, a significant portion of ground-level emissions originates from other sources. These primarily include the aircraft's Auxiliary Power Unit (APU) and ground service equipment (GSE).

The APU is a small turbine engine in the aircraft's tail that provides power and air conditioning while the main engines are switched off. As there are no international certification standards, information on APU emissions rates is not made available by the manufacturers. ICAO (2015) estimates that typical APU usage during aircraft turnaround consumes approximately 80 kg of fuel

Table 4.1. ICAO Reference Landing and Take-Off (LTO) Cycle.

Operating phase	*Time in mode (minutes)*	*Thrust setting (% of rated thrust)*
Take-off	0.7	100
Climb	2.2	85
Approach	4.0	30
Taxi and ground idle	26.0 (7.0 in + 19.0 out)	7

for a short-haul aircraft and up to 300 kg for a long-haul aircraft. This corresponds to around 700 g and 2,400 g of NO_x emissions, respectively and approximately 250 kg and 950 kg of CO_2. However, these numbers are estimates based on averaged consumption data. The specific APU operations depend strongly on the handling characteristics of the respective airports.

Additionally, the fleet of GSE vehicles (which includes baggage tugs, belt loaders, pushback tractors and catering trucks, often powered by diesel engines) contributes to local air pollution. Again, the specific emission volumes depend heavily on the operation of an airport. ICAO (2015) provides a calculation example for Zurich Airport, indicating that the turnaround handling emits approximately 260 g of NO_x (and ca. 20 kg CO_2) for a narrow-body aircraft and about 510 g of NO_x (ca. 48 kg CO_2) for a wide-body aircraft.[27]

Attributing Air Pollution to Aviation Sources

Assessing aviation's specific impact on local air quality presents a significant scientific challenge. The air quality is continuously monitored at permanent stations, and the concentration of air pollutants is determined. The stations register the total concentration of pollutants across all sources. They cannot, by themselves, distinguish between emissions originating from aircraft and those from other significant sources such as airport ground vehicles, nearby highway traffic or local industry. This problem of attributing any given deterioration in air quality to emitting sources is further complicated by the fact that monitoring stations are often deliberately placed in nearby residential areas to measure population exposure, rather than directly at the source of airport emissions.

To overcome this challenge and apportion the sources of pollution, scientists and regulators rely on sophisticated emission dispersion models. These models use detailed input data to estimate how pollutants from specific airport sources are transported, dispersed by wind and modified by chemical reactions in the atmosphere. Key inputs for these models include detailed emission inventories, flight track data and local meteorological conditions. By accounting for background pollution, these models try to estimate the specific, incremental impact of aviation on the air quality in surrounding communities. ICAO provides detailed guidance on this established methodology in its Airport Air Quality Manual.[28]

While specialised models are necessary for a quantitative assessment at a given location, it remains a fundamental principle that pollutant concentrations are highest in the immediate vicinity of their source and are directly correlated with the scope of airport operations.

Quantifying the Impact of Aviation on Air Quality

While at many airports local air quality standards are clearly met at the monitoring stations,[29] the specific contribution of aviation remains indistinguishable from other sources. A report published by the Concawe Air

Quality Management Group (2023) highlights the significant local influence of aviation. Using models for six major European airport regions, the study found that while aviation's contribution to pollution in city centres is small (for example 2.5% for NO_2), it is significant directly at the airport locations, accounting for 38% of NO_2 and 45% of SO_2 concentrations. The analysis also showed that this impact diminishes rapidly with increasing distance from the airport.

The difficulty in measuring the effects of aviation on local air quality is further illustrated by conflicting findings from studies on the 2010 Eyjafjallajökull volcanic eruption, which grounded most European air traffic. An ACI (2010) report concluded that the airport closures had a very small effect on local air quality, suggesting other factors like wind were more dominant. However, research by Carlslaw, Williams and Barratt (2012) using different methodologies demonstrated an unambiguous decrease in NO_x concentrations attributable to the flight ban. This shows that while direct measurement is challenging, the estimated impact on local air quality depends strongly on the modelling techniques applied.

The Persistent Challenge of NO_x Emissions

Whereas overall fuel efficiency is improving, cutting NO_x emissions remains more difficult than reducing CO_2. Data from EASA (2025a) shows a clear decoupling of these trends. For instance, fuel burn per passenger-kilometre (pkm) in Europe decreased by 32% between 2005 and 2023 and is projected to fall by a total of 47% by 2050, relative to the 2005 baseline. This significant efficiency gain leads to forecasts of a relatively modest net increase in European CO_2 emissions, which are projected to be 13% higher in 2050 compared to 2005, despite strong traffic growth.[30] In contrast, total NO_x emissions from flights departing at European airports were 35% higher in 2023 than in 2005 and are forecasted to increase by approximately 88% by 2050.

This divergence occurs because reducing CO_2 is a direct function of improving fuel efficiency, whereas mitigating NO_x requires technologically more challenging improvements to the engine's high-temperature combustion process. Furthermore, the current regulatory focus, market-based measures and the use of Sustainable Aviation Fuel (SAF) primarily target the reduction of the CO_2 footprint. As a result, aviation's share of total NO_x emissions from the transport sector in Europe has risen significantly over the past decades.[31]

From Local Air Quality to Global Health Impacts

It is crucial to note, however, that the impact of aviation on air quality is not solely a local problem resulting from the emissions during the LTO cycle. Scientific studies have shown that emissions at cruise altitude that are descending to the ground are, in fact, the dominant contributor to the overall health impacts from aviation's particulate matter. Barrett, Britter and Waitz (2010) estimate that the emissions of $PM_{2.5}$ during the cruise phase of flights result in 80% of all aviation-related premature mortality, and Grobler et al. (2019) estimate that the global aviation $PM_{2.5}$ emissions lead to approximately 16,000 premature deaths annually.[32]

Climate Change and Aviation

While noise and air pollution are primarily local issues, aviation's impact on the global climate is its most significant and complex long-term environmental challenge. As a problem resulting from the global economy's reliance on fossil fuels, climate change is the subject of extensive international debates. Within this context, aviation's role is often not fully understood. The focus typically lies on CO_2 with aviation's direct share of global human-induced CO_2 emissions being around 2.5% in 2023, representing 941 million tonnes.[33]

Data on Aviation's CO_2 Emissions

The International Energy Agency (IEA) data quoted above includes all aircraft emissions, i.e., on domestic and international flights for civil as well as for military purposes. For commercial air transport, IATA (2025) reports 882 million tonnes of CO_2 emissions in 2023, leading to a share of almost 94% of aviation's emissions stemming from commercial aviation. Other studies find lower values for other years, e.g., around 88% of aviation's CO_2 emissions were attributed to commercial aviation in 2017; accordingly, the share of military aircraft was about 8%, another 4% of CO_2 emissions stemming from private flights.[34] For 2024, IATA (2025) reports 942 million tonnes of CO_2 emissions from commercial aviation, an increase by 6.8% against 2023, and the highest quantity of emissions recorded so far.

In theory, three options exist for allocating these emissions to different countries and world regions. First, emissions might be allocated based on the home country of the airline causing the respective emissions. Second, emissions on all outgoing flights from one country might be attributed to the respective country (this to a large degree covers the emissions from aviation fuel sold in the respective country). Third, one might attribute emissions to the home country of the respective passengers (and shippers in the case of air cargo), which for practical implication requires the largest amount of data.

However, the industry's actual contribution to global warming is disproportionately larger due to a range of non-CO_2 effects that are unique to emissions at high altitudes. The most recent and comprehensive assessments by authorities like ICAO, EASA or the Intergovernmental Panel on Climate Change (IPCC) now conclude that these non-CO_2 effects are responsible for approximately two-thirds of aviation's total net warming effect (see below).

Fundamentals of the Greenhouse Effect and Climate Change

Climate change is a global challenge, best described as a general increase of the Earth's average temperature, a phenomenon also known as global warming. This warming is induced by the greenhouse effect, a natural process essential for life on earth. Certain atmospheric components, known as greenhouse gases (GHG), allow short-wave solar radiation to pass through but absorb a portion of the long-wave thermal radiation emitted back from the Earth's surface, trapping heat much like the glass of a greenhouse. Key GHGs include H_2O, CO_2, methane (CH_4) and NO_x.[35] While natural processes (like biological processes, volcanic eruptions, etc.) produce these GHG, human activities since the industrial revolution (and using fossil fuel resources) have further increased the concentration of these gases in the atmosphere leading to an enhanced greenhouse effect. The fundamental correlation between GHG concentrations and global temperatures is well-established (see Figure 4.4 which depicts the atmospheric CO_2 concentration (measured in parts per million, ppm) and the global mean surface temperature anomaly (expressed as index in °C relative to the 1951–1980 baseline period) over the time span

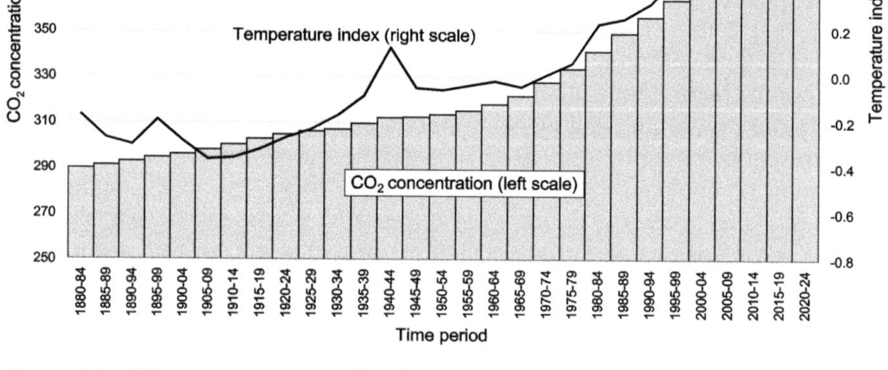

Figure 4.4. CO_2 concentration and global temperature index from 1880–2024. Figure created using data from NASA (2025), Meinshausen et al. (2017) and NOAA (2025).

1880–2024). However, the exact interactions, feedback loops and microphysical processes within the climate system remain a focus of intensive scientific research worldwide.

Consequences of Global Warming

Global warming has several critical and far-reaching consequences that define climate change. One of the most visible consequences is the melting of snow fields and glaciers. This process contributes directly to sea-level rise, but also triggers a feedback loop: as white, reflective surfaces of ice and snow disappear, the darker land and ocean surfaces underneath absorb more solar energy, further accelerating warming. Another key consequence is the intensification of the hydrological cycle. A warmer atmosphere can hold more moisture,[36] leading to more extreme precipitation events and severe flooding in some regions, while other areas experience longer and more intense droughts. Furthermore, global warming alters large-scale atmospheric circulation patterns. Polar regions warm faster than tropical regions. This reduction in temperature difference destabilises the jet stream, a key driver of weather in the mid-latitudes. A slower meandering jet stream can cause weather systems to become "stuck", leading to prolonged heatwaves, persistent droughts or continuous rainfall over the same areas for extended periods.

It is important for the aviation industry to recognise that it is not only a contributor to these effects, but also highly vulnerable to them. Rising sea levels directly threaten coastal airports, while more intense and frequent extreme weather events disrupt flight operations, causing delays, cancellations and consequently significant economic losses.[37]

Sources of Aviation's CO_2 Emissions

As already stated above, the combustion of hydrocarbon-based jet fuel is the primary source of aviation's CO_2 emissions. When assessing the industry's total carbon footprint, it is crucial to distinguish between emissions of the two main contributors: airlines (by operating aircraft) and airports (by providing the necessary ground infrastructure). While the operation of an aircraft is usually fully controlled by the airline, emissions from airport operations can be generated directly by the airport or indirectly from the airport's partners and customers. The framework for this "allocation" of emissions and also responsibilities is the GHG Protocol, which categorises emissions into three scopes (see also Chapters 7 and 8).[38] For a typical airport, the emissions under its direct control – Scope 1 (e.g., from heating plants and airport-owned vehicles) and Scope 2 (from purchased electricity) – represent only a small fraction of the total emissions. The overwhelming majority of emissions falls under Scope 3, which includes all other indirect emissions. As discussed regarding local air pollution, this category is dominated by the aircraft operations themselves (LTO cycle) but also includes emissions from passenger and staff ground transport to and from the airport. While precise figures vary by airport,[39] industry reports consistently show that Scope 1 and 2 emissions typically account for less than 5% of the total carbon footprint associated with an airport.[40]

The High-Altitude Problem: A Stable Environment

The unique significance of aviation's non-CO_2 effects results from the fact that aircraft emit substances directly into the upper atmosphere, at typical cruise altitudes of 8–13 kilometres. This region, encompassing the upper troposphere and the lower stratosphere, is characterised by its high stability, with limited vertical air movement and mixing.[41] As a result, substances released here, including GHG and PM, have a much longer atmospheric residence time compared to emissions at lower altitudes. This prolonged persistence means their potential to alter atmospheric chemistry and the Earth's energy balance is significantly amplified.[42]

Water Vapour, Contrails and Aviation-Induced Cloudiness

The direct emission of the GHG H_2O also contributes to climate change. While H_2O emissions in the lower, moist troposphere are negligible, water vapour injected directly into the typically very dry stratosphere has a small but direct warming effect, as it increases the GHG concentration in this particularly sensitive atmospheric region.

However, the most significant climate impact related to water vapour is the formation of condensation trails (contrails). In the extremely cold (typically below -40 °C) and humid conditions of the upper atmosphere, water vapour emitted by engines rapidly condenses and freezes around condensation nuclei, primarily soot particles. While many of these ice-crystal contrails dissipate quickly, those that form in specific atmospheric areas known as Ice Super Saturated Regions (ISSRs) can persist for hours. Here, they spread out due to wind shear and evolve into extensive cirrus clouds. This phenomenon is termed Aviation-Induced Cloudiness (AIC).

These "artificial" clouds have a dual effect on the Earth's energy balance. During the day, they reflect some incoming solar radiation back to space (a cooling effect) while also trapping outgoing thermal radiation from the Earth (a warming effect). At night, however, there is no incoming sunlight to reflect, so they only trap heat, resulting in a net warming effect, which is particularly relevant for flights occurring at night.[43]

Atmospheric Chemistry of NO_x and SO_x

Beyond cloud formation, emissions at high altitudes also trigger complex chemical reactions. NO_x, which is a local air pollutant at ground level, has a different impact at cruise altitude. Here, in the presence of intense sunlight,

NO_x acts as a catalyst in the production of ozone (O_3), a powerful but short-lived GHG. But NO_x also leads to the destruction of existing CH_4, a long-lived and powerful GHG, which results in a cooling effect. Unfortunately, the net result of all these opposing forces is a significant net warming effect. SO_x, coming from sulphur in the fuel, form sulphate aerosols. These particles have a minor direct cooling effect by reflecting sunlight but serve as condensation nuclei for contrail formation.[44]

Assessing Aviation's Climate Effect: From RFI to Effective Radiative Forcing

To compare the impact of different climate drivers, scientists usually use the concept of Radiative Forcing (RF). It is the fundamental metric for a change in the Earth's energy balance, measured in Watts per square metre (W/m^2), quantifying how much a specific factor warms (positive RF) or cools (negative RF) the planet.[45] However, to better assess (and communicate) the impact aviation with its non-CO_2 effects has, scientists now primarily use the metric of Effective Radiative Forcing (ERF). While traditional RF calculates a climate driver's impact by keeping all atmospheric conditions fixed, the more advanced ERF accounts for rapid atmospheric adjustments coming from the different lifetimes of the various forcing agents. According to the IPCC (2013) this updated method provides a better prediction of the eventual temperature change and is now used for major drivers like well-mixed greenhouse gases and aerosols.

Aviation's Contribution to Climate Impact

Figure 4.5 presents an excerpt from the latest assessment of the climate impact of historic air traffic emissions between 1940 and 2018.[46] The left four columns show the ERF of major emission types, with the current best estimate for each indicated in the centre and the 5th and 95th percentiles at the lower and upper ends, respectively.

The best estimate for the ERF of CO_2 is 34 mW/m^2, with a relatively narrow confidence range between 28 and 40 mW/m^2. In contrast, the ERF for contrail cirrus is estimated at 57 mW/m^2, but with a much wider range (17 to 98 mW/m^2). Overall, as shown in the two rightmost columns, non-CO_2 effects account for approximately two-thirds of total radiative forcing. However, the confidence in these estimates remains low, primarily due to the substantial uncertainties in quantifying the contrail cirrus effect.

As mentioned above, according to the IEA (2024), aviation accounted for 2.5% of all human-made CO_2 emissions in 2023, a level consistent with the pre-pandemic year of 2019. Within the transportation sector, aviation is responsible for 12% of direct CO_2 emissions. For context, the transport sector as a whole contributes 23% of all global energy-related CO_2 emissions.[47]

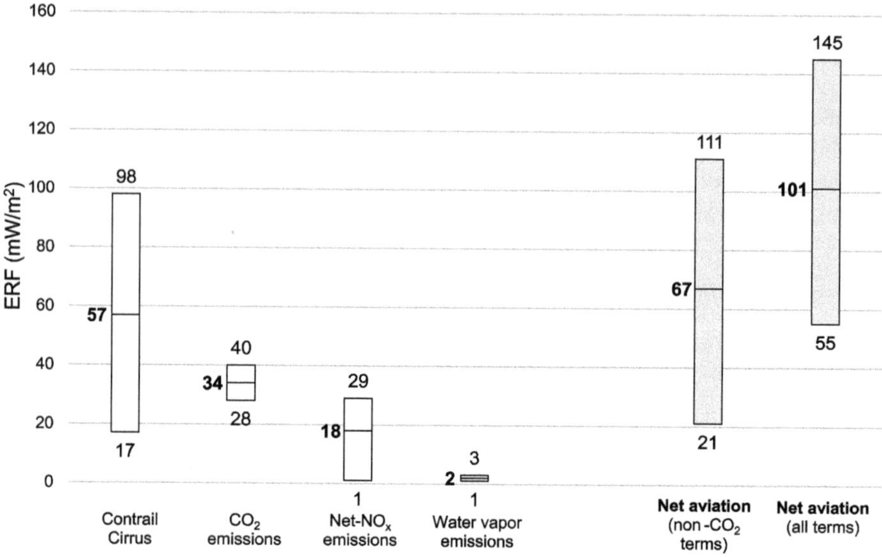

Figure 4.5. Latest best estimates for climate forcing terms of historic global aviation emissions from 1940 to 2018. Based on Lee et al. (2021).

A detailed analysis for 2019 by the ICCT (2020) provides a clearer picture of where these emissions originate, analysing a total of 920 Mt of CO_2 emissions stemming from commercial aviation. Passenger transport is the dominant source, accounting for approximately 85% of commercial aviation's CO_2 output, freight carriage is responsible for the remaining 15% (with 8% from belly freight carriage on passenger aircraft and 7% from dedicated freighter operations). The 85% of passenger transport can be broken down further by aircraft type: narrow-body jets contributed 43% to the total, wide-body jets 37% and regional aircraft around 6%. Geographically, five countries were responsible for nearly half of all passenger-related carbon emissions, based on the departure airports of the respective flights: USA (23%), China (13%), the United Kingdom (4%), Japan (3%) and Germany (3%). As a bloc, the European Union including the United Kingdom ranked second globally with 19% of the total. The data also shows that emissions were evenly distributed across flight distances, with roughly one-third each originating from short-haul (<1,500 km), medium-haul (1,500–4,000 km) and long-haul (>4,000 km) flights. To make these numbers from ICCT (2020) more tangible: in 2019 average CO_2 emissions per passenger were at 124 kg.[48] It is important to note that specific emissions vary with stage length: on short-haul flights, the energy-intensive take-off and climb phases have a greater impact, whereas on long-haul flights where the emissions during the LTO cycle are less significant the high fuel load increases overall emissions due to the additional weight.

The Challenge Beyond Carbon Neutrality

Aviation's climate impact, particularly the dominance of non-CO_2 effects, has profound implications for the industry. It demonstrates that achieving "climate neutrality" requires far more than achieving "carbon neutrality". Strategies that focus solely on CO_2 emissions, such as carbon offsetting or even the deployment of SAF, are insufficient to address the full warming impact of flying. The industry therefore faces a dual challenge: it must not only pursue the ambitious goal of decarbonising its energy supply but also simultaneously develop and implement new technologies and operational procedures to mitigate the complex, location-based non-CO_2 effects like AIC (see Chapter 6).

Other Environmental Aspects

Beyond the primary environmental impacts of noise, local pollution and climate change, a complete assessment of aviation's footprint should also consider other relevant aspects. These include the contamination of local water and soil from operational activities, waste generation and disposal as well as the ecological consequences of land use. While interconnected with the entire aviation system, these particular impacts are primarily managed at the airport level.

Water and Land Contamination

Beyond gaseous emissions that affect air quality and climate, aviation also produces liquid and solid pollutants at the airport level that pose a direct risk to water and soil. While minor spills of fuel or oil are typically contained by professional ground handling procedures, more significant contamination risks come from two main sources: the widespread application of aircraft de-icing and anti-icing fluids in winter operations and the deposition of tyre and brake wear particles.[49]

Of these sources, the runoff from aircraft de-icing is often reported as the most significant in terms of volume and immediate ecological impact. The glycol-based fluids are essential for safety, but their biodegradation in water consumes large amounts of dissolved oxygen, which can harm or kill aquatic life. In contrast, the impact of tyre and brake wear receives less attention as a source of microplastics and other pollutants also found in airport runoff.

De- and Anti-Icing Operations and Their Environmental Impact

During cold weather operations, the accumulation of even a thin layer of snow or ice on aircraft surfaces poses a significant safety risk as it can lead to a critical loss of lift and increased drag, particularly during take-off. For this reason, safety regulations worldwide mandate that aircraft surfaces must be completely clean before departure. This is achieved through a two-stage process: "de-icing" removes existing contamination, while "anti-icing" applies a protective layer to prevent new ice from forming before take-off. As salt or mechanical scratching would damage the sensitive aircraft structures, this process is conducted by spraying the aircraft with specialised, glycol-based fluids.[50]

The primary environmental problem results from the chemical properties of these de-icing fluids. When the fluid runs off the paved surfaces of the airport, it enters the local water system. Glycols are biodegradable, and microorganisms in the water consume the glycol rapidly. This process consumes large amounts of dissolved oxygen, reducing the oxygen level in (ground) water that is necessary for other organisms' lives, leading to the deaths of fish and other creatures.[51] Furthermore, while the glycols themselves are well understood, there is less comprehensive data on the long-term environmental effects of the various proprietary additives (such as corrosion inhibitors, thickeners and dyes) that are part of these de- and anti-ice liquids.

The volumes of fluid used can be substantial and reach up to 15,000 litres per aircraft in severe weather.[52] While airports invest in dedicated de-icing pads with collection systems, a significant portion of the fluid (particularly the thicker anti-icing fluid designed to stick to the aircraft) is sloughed off during taxi and take-off, ending up on and adjacent to the runways. Estimates suggest that 15–20% of the applied fluid is lost in this way.[53] According to Sulej, Polkowska and Namiesnik (2012), the U.S. Environmental Protection Agency (EPA) once estimated that this results in as much as 40 million litres of de-icing fluid being discharged annually into US waterways, highlighting the scale of the challenge.

Tyre and Brake Wear: An Emerging Environmental Issue

An often-overlooked environmental impact comes from the abrasion of aircraft tyres and brakes during landing and taxiing. While this topic has received less attention than other impacts, recent research highlights it as an emerging threat, particularly to soil health.[54] The most intense moment of abrasion occurs at touchdown. The aircraft's wheels are forced to accelerate from zero to the aircraft's landing speed (which can exceed 250 km/h) in a fraction of a second, generating immense friction and heat and resulting in a visible puff of smoke and the abrasion of rubber particles. Additional wear occurs during braking and taxiing manoeuvres.

The environmental problem is twofold. First, the abraded particles are a significant source of microplastics in the environment, and second, the mixture of its components can leach hazardous substances into the soil and adjacent water systems. These contaminants can harm soil microorganisms and biodiversity and have the potential to enter the food chain.

Quantifying the exact mass of rubber shed per landing is challenging, and available estimates vary widely depending on factors like aircraft weight, landing speed and runway conditions. While runway deposits must be regularly removed for operational safety, finer particles can be dispersed into the surrounding environment. Estimates for the amount of rubber deposited per tyre per landing range from tens of grams to several hundred grams.[55] This highlights that, while the individual amounts are small, the cumulative effect of thousands of landings at a major airport constitutes a continuous source of these specific pollutants into the local ecosystem.

Waste Generated by Aviation

Waste management presents another logistical and environmental challenge for the aviation industry. Airports and airlines generate substantial and diverse waste types and volumes, from industrial waste during maintenance to municipal-style waste in terminals and cabin waste from flights. To quantify the scale of the issue, for example ATAG (2024) reports from an IATA study, every passenger produces on average 762 g of waste (almost 1.18 kg on long-haul flights). In 2023, 3.6 million tonnes of cabin waste were generated, and waste removal costs the airlines about 6 billion USD per year.[56]

The management of this waste[57] is defined by a crucial distinction between waste generated on the ground and waste from aircraft cabins. Waste from airport terminals, shops, and restaurants is typically managed in line with local and national regulations, similar to any other large commercial facility, with established systems for sorting and recycling. The same applies to specialised streams like construction and demolition debris or industrial hazardous waste from maintenance operations.

The unique and complex challenge for aviation, however, lies with cabin waste from international flights. To prevent the cross-border transmission of animal and plant diseases, many countries (especially those with economically important agricultural sectors) have strict biosecurity regulations that classify this as a high-risk material. These regulations often legally mandate that the entire, mixed waste stream from an international flight (including otherwise clean and recyclable materials) must be sent directly to high-temperature incineration or deep-burial landfill. This severely limits recycling opportunities and represents a core challenge for improving the industry's circularity.

Land Use and Its Ecological Impacts

The physical footprint of aviation in the form of land use is a significant, direct and often permanent environmental impact. Airports are among the largest single pieces of infrastructure in any region, with major international hubs occupying large areas of land.[58] This land is required not just for the core operational elements like runways, taxiways and terminals, but also for a host of ancillary facilities, including cargo terminals, maintenance hangars, fuel farms and especially also legally required safety zones.

This extensive land use has several direct ecological consequences. The construction of airport infrastructure leads to land sealing with concrete and asphalt, which prevents rain infiltration and impacts soil ecosystems. The most significant effect is often the habitat loss for local flora and fauna, as airports are frequently built on previously natural or agricultural land. Furthermore, the vast, fenced-off airport area acts as a physical barrier, leading to ecosystem fragmentation.

While the absolute area of an airport is large, a more detailed perspective is required. First, because an airport is a single, defined area, its effect as a physical barrier is usually limited, especially when compared to linear infrastructure like highways and railway tracks. Second, a significant portion of an airport's total area often consists of unsealed green spaces between runways and taxiways. At German airports, for example, around 80% of the total ground consists of unsealed areas.[59] These extensive, largely undisturbed areas can become valuable refuges for specific flora and fauna that are exposed to threats elsewhere. And finally, when compared to the vast networks required for ground-based transport, aviation's total land use is very small. In Germany, for instance, aviation infrastructure accounts for only around 2% of the total area dedicated to the transport sector, whereas road transport is responsible for the vast majority.[60]

Another aspect of aviation's impact on land use is linked to the production of biogenic Sustainable Aviation Fuel or biofuel (see Chapter 8). While biogenic SAF use is based on the principle that carbon absorbed during biomass growth is released again during combustion, the actual GHG emission reductions depend on how and where the biomass feedstock is produced. Biomass feedstocks can consist of waste and residues but also of purpose-grown energy crops that require land and therefore compete with other land uses. Direct land use change (DLUC) occurs when land is explicitly converted for SAF feedstock production, whereas indirect land use change (ILUC) arises when agricultural land previously used for food or feed is diverted to SAF production and the displaced production shifts to non-agricultural land such as forests or grasslands.[61] Both DLUC and ILUC can partially or even fully offset the GHG savings achieved by replacing fossil jet fuel. Empirical and model-based studies show that resulting emission reductions vary widely across SAF pathways, with ILUC emissions representing one of the main sources of uncertainty in life cycle assessments, as these effects are not directly observable and must be estimated.[62] In general, waste- and residue-based SAF pathways are considered to involve low DLUC or ILUC emissions and therefore offer robust emission reduction potential, whereas crop-based pathways exhibit substantial variability depending on feedstock type, regional context and soil carbon dynamics.

Monetarisation of Environmental Damage

In economics, the detrimental effects on the environment are sometimes monetarised, allowing for a comparison of the share of different effects from a given industry or activity as well as for intermodal comparisons, e.g., average external costs per passenger-kilometre (pkm) from aviation and high-speed rail. However, in order to come up with such a unidimensional metric, it has to be assessed how specific emissions cause damage to the environmental, how damage can be attributed to different sources, and which "price tag" should be put on a specific type of damage. One of the most challenging areas is climate change, because its effects are global (leading to the question whether different or uniform costs should be used for damage occurring in different countries) as well as long-lasting (leading to the question whether damage that affects future generations should be discounted or not). Moreover, the valuation of a human's health and life will always remain a controversial issue.

For the European Union, the "Handbook on the external costs of transport" summarises methodologies as well as results.[63] This Handbook not only refers to environmental externalities but also covers accidents and congestion, which from an economic perspective can be seen as externalities (see also Chapter 6). For the passenger segment, average external costs per air transport passenger-kilometre (pkm)[64] in 2016 are calculated at a level of 3.4 euro cents, which is slightly more than for rail (2.8 euro cents), but significantly less than for cars (12.0 euro cents).[65] The largest share in aviation's external costs are the costs of climate change (2.2 euro cents = 65%), followed by "well-to-tank" costs, i.e., external costs associated with fuel productions and transport (0.9 euro cents = 26%). Air pollution causes external costs of 0.2 euro cents per pkm.

Key Takeaways

- Aviation's environmental impacts span local, regional and global scales and have evolved over time, with public and regulatory attention shifting from noise and local pollution towards climate-related effects.
- Aircraft noise per movement has been significantly reduced through technological progress, but continued traffic growth offsets these gains and keeps overall noise exposure at a high level.
- Local air quality around airports remains a persistent environmental challenge, as emissions are spatially concentrated and particularly NO_x reductions depend on combustion technologies rather than fuel efficiency alone.
- Despite efficiency improvements, aviation's share of transport-related NO_x emissions is increasing, highlighting the limits of fuel burn reductions as a sole mitigation strategy.
- Climate change has emerged as aviation's most critical environmental issue, with the sector contributing around 2.5% of global anthropogenic CO_2 emissions.
- Non-CO_2 effects, especially contrail cirrus formation, dominate aviation's total warming impact, yet remain subject to substantial scientific uncertainty, complicating mitigation measures and regulatory responses.

- Additional environmental impacts, including water contamination from de-icing fluids, microplastic emissions from tyre and brake wear and increasing waste streams, further add to aviation's overall environmental footprint.
- Upstream emissions from fuel production, refining and transport constitute a significant share of aviation's life cycle climate impact and must be considered in comprehensive environmental assessments.

Notes

1 An exception to this is the case of supersonic flight. For example, the Concorde generated loud sonic booms during cruise flight, which led to restrictions on supersonic overland flight in many countries. As a consequence of this, the mitigation of sonic boom effects remains a central focus in the research and development of next-generation supersonic aircraft.
2 CANSO/ACI (2015).
3 The physiological consequences of (chronic) noise exposure have been subject of research for many years and in many disciplines. In their "Environmental Noise Guidelines" the WHO provides a comprehensive overview about the different studies and their outcomes – all confirming the consequences described here.
4 This is confirmed for example by the so-called RANCH study (Road traffic and Aircraft Noise and children's Cognition and Health study). It was a major pan-European research project that provided key evidence on the effects of chronic exposure to transport noise on children's cognitive development, particularly reading comprehension (Stansfeld et al. 2005).
5 For example, Kopsch (2016) provides a meta-analysis about the reduction in housing values due to aircraft noise: for every one-decibel (dB) increase in the average noise level, housing values decrease on average by 0.68%.
6 For example, a human would be able to notice the rise in the physical sound energy of dropping one or two pins, but not between dropping 1,000 or 1,001 pins although the absolute sound energy difference of one dropping pin is the same in both situations. On this scale, a doubling of physical sound energy would not correspond to a doubling in dB but an increase by 3 dB (for example, two 50 dB events would result in 53 dB – not 100 dB).
7 More (physical) details of these and also additional metrics are described in the "ERCD REPORT 0904 Metrics for Aircraft Noise" from CAA (2009).
8 For example, compared to an aircraft producing a high peak noise level for a short duration another aircraft with a lower peak but longer duration might have a higher total sound energy and thus a higher SEL value and is potentially being perceived as "noisier" overall.
9 Due to the logarithmic nature of the decibel scale, a 10-dB penalty means that a single noise event at night is weighted in the DNL calculation as if it were 10 identical events occurring during the day (FAA 2025a).
10 Examples are provided in CAA (2014).
11 This EASA analysis refers to 98 major airports in the EU27+EFTA region, representing approximately 86% of the total acoustic energy from LTO operations in 2023.
12 This refers to the data from the 4th reporting round of the Environmental Noise Directive (END 2002/49/EC), which covers approximately 50 airports in the EU27 +EFTA countries for the reference year 2021.
13 When applying the stricter WHO guideline thresholds to the same 2021 data, these figures increase substantially: the number of highly annoyed people rises to 2.3 million and the highly sleep disturbed to 680,000, with 14.9 million people considered exposed during the day and 5.3 million at night.

14 See EASA (2025a). If the health impact rates from the low-traffic year 2021 (for example ca. 127,000 highly sleep-disturbed people from 0.7 million exposed at night) were extrapolated to the pre-pandemic 2019 exposure levels (1.5 million exposed at night), the number of affected people would more than double.
15 BTS (2025).
16 This is confirmed by many studies, see for example the reports from WHO (2018) or EEA (2025a) but also scientific publications, e.g., Stansfeld et al. (2005) or Kopsch (2016).
17 For example, despite being larger and heavier, the modern Boeing 747–8 has a cumulative noise footprint around 36 dB (EPNL) less than its original predecessor Boeing 747–100, see EASA (2025c).
18 ATAG (2024).
19 The Air Transport Action Group (ATAG) estimated that in the year 2023, approximately 11.6 million people were directly employed by the aviation industry (at airlines, airports, air navigation services and in civil aerospace manufacturing) with more than 50% working for airports or on-airport positions. A significant portion of these employees, particularly those working airside, experience higher exposure to emissions than the general public (ATAG, 2024).
20 IPCC (1999).
21 Eurocontrol (2018).
22 ICAO (2024).
23 Many reports from organisations like ICAO, EASA, IPCC etc. contain further details about these pollutants.
24 Particulate matter is categorised by its diameter. PM_{10} refers to particles smaller than 10 micrometres (μm), $PM_{2.5}$ refers to fine particles smaller than 2.5 μm and Ultrafine Particles (UFP), also known as $PM_{0.1}$, are smaller than 0.1 μm (or 100 nanometres).
25 For example, during a typical LTO cycle, an Airbus A320 is estimated to emit around 2.4 tonnes of CO_2 and 9 kg of NO_x. In comparison, an Airbus A350 emits approximately 6.7 tonnes of CO_2 and 40 kg of NO_x, while an Airbus A380 reaches about 12.6 tonnes of CO_2 and 69 kg of NO_x (ICAO 2015).
26 ICAO (2015) and TRB (2014).
27 To put these figures into perspective, an illustrative example for a typical narrow-body aircraft can be used. During the LTO cycle the aircraft emits around 9 kg of NO_x (and ca. 2.5 t CO_2). During a typical turnaround at the gate, an additional 700 g of NO_x (ca. 250 kg CO_2) is emitted by the APU and about 260 g of NO_x (ca. 20 kg CO_2) by ground service equipment (GSE). This underlines that direct aircraft emissions are by far the dominant source even when considering only the turnaround phase with inoperative main engines. The contribution from GSE amounts to merely 2.5% of total NO_x emissions (260 g of 9,700 g) and less than 1% of total CO_2 emissions (20 kg of 2,750 kg).
28 ICAO (2015).
29 For example, Frankfurt Airport (Fraport 2022) reported in 2022 that at its 15 monitoring stations the annual mean values of pollution components were all within air quality standards.
30 All these projections are based on EASA's "baseline" forecast scenario, which assumes an average development of traffic growth and the introduction of new technology and operational improvements.
31 See EASA (2025a). According to Krecl et al. (2021), road transport remains the main anthropogenic source of NO_x in Europe, but mitigation policies have been highly effective in lowering on-road vehicle emissions. EEA (2025b) reports a reduction of more than 50% in NO_x emissions across EU member states between 2005 and 2023.
32 To put this number into context, the WHO's (2024) latest fact sheet estimates that air pollution from all sources combined causes around 6.7 million premature deaths globally each year.

33 IEA (2024).
34 Gössling and Humpe (2020).
35 Without this natural effect, the average global temperature would be an inhospitable -18 °C.
36 Actually, this leads again to a feedback loop: a warmer atmosphere can hold more moisture, and as water vapour itself is a GHG, this leads to further warming.
37 See EASA (2025a) for reports and strategies how the aviation industry could adapt to the impacts of climate change.
38 EASA (2025a).
39 For a case study, summing up all CO_2 emissions from travellers using a regional airport in Germany, see Klophaus and Lauth (2022).
40 With the exception of the FAA (2021), which reports Scope 1 and 2 emissions of less than 2%, industry-wide figures remain difficult to obtain. The FAA estimation is confirmed for European aviation by own calculations. Furthermore, these figures do not include the non-CO_2 warming effects, which are generated exclusively by the aircraft during flight and, as detailed in the following sections, constitute the majority of aviation's total climate impact, thus, the dominance of the aircraft itself as the primary source of the total climate impact becomes even more pronounced.
41 The stability of the stratosphere is caused by its unique temperature profile. Unlike the troposphere below, where temperature decreases with altitude (allowing for vertical convection and mixing), in the stratosphere temperature generally increases with altitude.
42 As one of many examples, the IPCC (1999) Special Report Aviation and the Global Atmosphere provides further details about the exact interactions of the non-CO_2 emissions.
43 Teoh et al. (2024) have estimated that a very small fraction of flights (only 2%) that fly through these specific atmospheric regions are responsible for up to 80% of the total contrail climate forcing. This finding is one example for ongoing exploration about AIC, that currently is one main focus in research about aviation's contribution to climate change.
44 The different chemical effects highlight the complexity of mitigation. For example, the use of certain Sustainable Aviation Fuels with a lower sulphur and aromatic content can reduce the formation of soot and sulphate aerosols. While this is beneficial for reducing contrail formation, the reduction of cooling sulphate aerosols could slightly offset some of the climate benefits, a trade-off that is also subject of ongoing research.
45 Another common metric is the Global Warming Potential (GWP), which compares the warming effect of a gas relative to CO_2 over a specific time horizon (typically 100 years). However, GWP is designed for well-mixed, long-lived gases and is of limited relevance for assessing the unique, short-lived and altitude-dependent climate effects of aviation like for example contrails.
46 Lee et al. (2021).
47 IPCC (2025).
48 90 g CO_2 per pkm, the average flight distance was 1,378 km.
49 Large-scale pollutant releases from emergencies, such as from firefighting or fuel dumping (jettisoning fuel to reduce landing weight during emergency landings), are extremely rare occurrences and are typically excluded from routine environmental assessments.
50 There are four main types of fluids (Type I, II, III, IV) standardised by SAE International. Type I fluids are thin, typically dyed orange, and used for the initial de-icing process. Type IV fluids, which are much thicker and often dyed green, are used for anti-icing, providing a protective layer that can last until take-off. Reports and Guidelines for Anti-Icing Operation like the one from Transport Canada (2021) and research papers like from Sulej, Polkowska and Namiesnik (2012) or

Switzenbaum et al. (2001) provide further details on the handling and consequences of the use of de- and anti-icing liquids.

51 Marais, Wolfe and Waitz (2016).

52 Sulej, Polkowska and Namiesnik (2012).

53 Switzenbaum et al. (2001).

54 Although research specifically on aircraft tyre abrasion is scarce, the environmental consequences of car tyre abrasion are well-documented (see for example Ding et al., 2022 and Federico et al., 2023). Given that there are no fundamental differences in material composition between car and aircraft tyres, and since abrasion from aircraft tyres is an undeniable phenomenon, the findings from these studies can be reasonably considered applicable to aircraft tyres as well.

55 Estimates range from approximately 45 g to higher estimates sometimes cited as several hundred grams (an often-quoted figure is 0.7 kg). However, given the variety in these numbers and the uncertainties and unavailability of the sources, these figures should be treated with caution.

56 According to Sebastian and Louis (2021) the COVID-19 pandemic temporarily altered these figures and compositions, leading to a surge in hazardous waste from Personal Protective Equipment (PPE) and a reduction in recycling due to contamination concerns.

57 ICAO's (2018) manual "Waste Management at Airports" provides further details.

58 For example, the largest airport of the USA, Denver International Airport has a total land area of over 135 km^2, followed by Dallas/Fort Worth International Airport with nearly 70 km^2. Europe's largest airport is Paris Charles de Gaulle Airport with 32 km^2. See Worldatlas (2025).

59 Own calculations based on UBA (2020a).

60 See Destatis (2025). However, this perspective can shift when land use is analysed not in absolute terms, but on a per-user basis (e.g., per pkm). Due to the methodological challenges of such allocations, data is scarce, but it provides an important alternative viewpoint. A study by the German Environment Agency (UBA) provides an example of such a user-based analysis for the year 2017. It estimates that for domestic flights, the land use per pkm is approximately 65% higher than for car traffic and about 30% lower than for long-distance rail. For international flights, which benefit from higher occupancy and longer distances, the land use per pkm is significantly lower and broadly comparable to other transport modes (UBA, 2020b). It is important to note that the per-user perspective could, in principle, be applied to other environmental impacts discussed in this chapter.

61 EASA (2025b).

62 Wang et al. (2025) and Zhao et al. (2021).

63 EC (2020).

64 The authors of the handbook point out that this number is based on selected larger airports and "may not be representative for all EU airports".

65 However, the environmental damage per pkm is almost as high as in aviation (3.3 euro cents), the largest elements of external costs are accidents (4.5 euro cents) and congestion (4.2 euro cents).

References

ACI Europe – Airports Council International Europe (2010) Effects of Air Traffic on Air Quality in the Vicinity of European Airports; Brussels; https://www.aci-europe.org/downloads/resources/ACI%20EUROPE%20Local%20Air%20Quality%20study%202010.pdf

ATAG – Air Transport Action Group (2024) Aviation Benefits Beyond Borders; https://aviationbenefits.org/downloads/aviation-benefits-beyond-borders-2024

Barrett, S.R.H., Britter, R.E., Waitz, I.A. (2010) Global Mortality Attributable to Aircraft Cruise Emissions; in: *Environmental Science & Technology*, Volume 44 (19); pp. 7736–7742; doi:10.1021/es101325r

BTS – Bureau of Transportation Statistics (2025) Number of People Residing in Areas of Significant Noise Exposure Around U.S. Airports; https://www.bts.gov/content/number-people-residing-areas-significant-noise-exposure-around-us-airports

CAA – Civil Aviation Authority (2009) Metrics for Aircraft Noise – Environmental Research and Consultancy Department (ERCD) Report 0904; https://www.caa.co.uk/publication/download/13689

CAA – Civil Aviation Authority (2014) Managing Aviation Noise (CAP 1165); https://www.caa.co.uk/publication/download/14877

CANSO/ACI – Civil Air Navigation Services Organisation and Airports Council International (2015) Managing the Impacts of Aviation Noise – A Guide for Airport Operators and Air Navigation Service Providers; https://store.aci.aero/product/aci-and-cansos-managing-the-impacts-of-aviation-noise-a-guide-for-airport-operators-and-air-navigation-service-providers-2015/

Carlslaw, D.C., Williams, M.L., Barratt, B. (2012) A short-term intervention study – Impact of airport closure due to the eruption of Eyjafjallajökull on near-field air quality; in: *Atmospheric Environment*, Volume 54; pp. 328–336; doi:10.1016/j.atmosenv.2012.02.020

Clark, C., Vienneau, D., Aasvang, G.M. (2025) Noise and Effects on Health and Well-Being; in: van Kamp, I., Woudenberg, F. (Eds.) *A Sound Approach to Noise and Health*; Springer, Singapore; pp. 51–73

Concawe (2023) The impact of aviation emissions to urban air quality in Europe – Detailed airport-city analysis; https://www.concawe.eu/publication/the-impact-of-aviation-emissions-to-urban-air-quality-in-europe-detailed-airport-city-analysis

Destatis (2025) Floor area total according to types of use in Germany; https://www.destatis.de/EN/Themes/Economic-Sectors-Enterprises/Agriculture-Forestry-Fisheries/Land-Use/Tables/areas-new.html

Ding, J., Lv, M., Zhu, D., Leifheit, E.F., Chen, Q.-L., Wang, Y.-Q., Chen, L.-X., Rillig, M.C., Zhu, Y.-G. (2022) Tire wear particles: An emerging threat to soil health; in: *Critical Reviews in Environmental Science and Technology*, Volume 53 (2); doi:10.1080/10643389.2022.2047581

EASA – European Union Aviation Safety Agency (2019) European Aviation Environmental Report 2019; https://www.easa.europa.eu/sites/default/files/eaer-downloads//219473_EASA_EAER_2019_WEB_HI-RES_190311.pdf

EASA – European Union Aviation Safety Agency (2025a) European Aviation Environmental Report 2025; https://www.eurocontrol.int/sites/default/files/2025-01/eurocontrol-easa-eaer-2025.pdf

EASA – European Union Aviation Safety Agency (2025b) How 'sustainable' are SAF?; https://www.easa.europa.eu/en/domains/environment/eaer/sustainable-aviation-fuels/how-sustainable-are-saf#saf-and-non-co2-emissions

EASA – European Union Aviation Safety Agency (2025c) EASA Certification Noise Levels; https://www.easa.europa.eu/en/domains/environment/easa-certification-noise-levels

EC – European Commission (2020) *Handbook on the external costs of transport*; Version 2019–1.1; Brussels; doi:10.2832/51388

EEA – European Environment Agency (2025a) Environmental noise in Europe – 2025; https://www.eea.europa.eu/en/analysis/publications/environmental-noise-in-europe-2025/environmental-noise-in-europe-2025
EEA – European Environment Agency (2025b) Trends in EU emissions of NH3, NMVOCs, NOx, SO2, primary PM2.5, primary PM10, BC and CO, between 2005 and 2023; https://www.eea.europa.eu/en/analysis/publications/air-pollution-in-europe-2025-reporting-status-under-the-national-emission-reduction-commitments-directive/trends-in-eu-emissions-of-nh3-nmvocs-nox-so2-primary-pm2-5-primary-pm10-bc-and-co-between-2005-and-2023
Eurocontrol (2018) European Aviation Fuel Burn and Emissions Inventory System for the European Environment Agency (for data from 2005); https://www.eurocontrol.int/publication/european-aviation-fuel-burn-and-emissions-inventory-system-feis-european-environment
FAA – Federal Aviation Administration (2021) United States Aviation Climate Action Plan; https://www.faa.gov/sites/faa.gov/files/2021-11/Aviation_Climate_Action_Plan.pdf
FAA – Federal Aviation Administration (2025a) Glossary; https://www.faa.gov/regulations_policies/policy_guidance/noise/glossary
FAA – Federal Aviation Administration (2025b) Fundamentals of Noise and Sound; https://www.faa.gov/regulations_policies/policy_guidance/noise/basics
Federico, L., Masseroni, A., Rizzi, C., Villa, S. (2023) Silent Contamination: The State of the Art, Knowledge Gaps, and a Preliminary Risk Assessment of Tire Particles in Urban Parks; in: *Toxics*, Volume 11 (5); 445; doi:10.3390/toxics11050445
Fraport (2022) Lufthygienischer Jahresbericht 2022; https://www.fraport.com/content/dam/fraport-company/images/umwelt/klimaschutz/luftqualit%C3%A4t/lufthygienische-jahresberichte/LHJB_2022_FRA_final.pdf/_jcr_content/renditions/original.media_file.download_attachment.file/LHJB_2022_FRA_final.pdf; available in German only
Gössling, S., Humpe, A. (2020) The global scale, distribution and growth of aviation: Implications for climate change; in: *Global Environmental Change*, Volume 65; 102194; doi:10.1016/j.gloenvcha.2020.102194
Grobler, C., Wolfe, P.J., Dasdhikari, K., Dedoussi, I.C., Allroggen, F., Speth, R.L., Eastham, S.D., Agarwal, A., Staples, M.D., Sabnis, J., Barrett, S.R.H. (2019) Marginal climate and air quality costs of aviation emissions; in: *Environmental Research Letters*, Volume 14 (11); 114031; doi:10.1088/1748-9326/ab4942
IATA – International Air Transport Association (2025) 2024 Aviation Emissions: Efficiency Gains vs. Rising Totals; https://www.iata.org/en/iata-repository/publications/economic-reports/2024-aviation-emissions-efficiency-gains-vs.-rising-totals
ICAO – International Civil Aviation Organization (2015) Airport Air Quality Manual; https://www.icao.int/sites/default/files/sp-files/environmental-protection/Documents/Doc%209889.SGAR.WG2.Initial%20Update.pdf
ICAO – International Civil Aviation Organization (2018): Waste Management at Airports – Eco Airport Toolkit; https://www.icao.int/sites/default/files/sp-files/environmental-protection/Documents/Waste_Management_at_Airports_booklet.pdf
ICAO – International Civil Aviation Organization (2022) 2022 Environmental Report; https://www.icao.int/sites/default/files/sp-files/environmental-protection/Documents/EnvironmentReport-2010/ICAO-ENV-Report-2022-F4.pdf
ICAO – International Civil Aviation Organization (2024) ICAO CAEP/12 Assessment Report: Assessment of the impact of airport emissions on local levels of NOx and human health, impacts of cruise emissions of NOx on human health, impacts of cruise NOx on climate; https://www.icao.int/sites/default/files/sp-files/environmental-protec

tion/Documents/ScientificUnderstanding/ICAO-CAEP12-Assessment-Report-on-impacts-of-aviation-NOx-emissions-on-air-quality-human-health-and-climate.pdf

ICCT – International Council on Clean Transportation (2020) CO_2 Emissions from Commercial Aviation 2013, 2018, and 2019; https://theicct.org/wp-content/uploads/2021/06/CO2-commercial-aviation-oct2020.pdf

IEA – International Energy Agency (2024) World Energy Outlook; https://iea.blob.core.windows.net/assets/140a0470-5b90-4922-a0e9-838b3ac6918c/WorldEnergyOutlook2024.pdf

IPCC – Intergovernmental Panel on Climate Change (1999) IPCC Special Report – Aviation and the Global Atmosphere; https://www.ipcc.ch/report/aviation-and-the-global-atmosphere-2

IPCC – Intergovernmental Panel on Climate Change (2013) Climate Change 2013: The Physical Science Basis; https://www.ipcc.ch/report/ar5/wg1

IPCC – Intergovernmental Panel on Climate Change (2025) IPCC Sixth Assessment Report, Working Group III: Mitigation of Climate Change, Chapter 10: Transport; https://www.ipcc.ch/report/ar6/wg3/downloads/report/IPCC_AR6_WGIII_Chapter10.pdf

Klophaus, R., Lauth, G.J. (2022) Monetary mapping of the climate footprint of air travel to a single airport; in: *Journal of Air Transport Management*, Volume 101; doi:10.1016/j.jairtraman.2022.102204

Kopsch, F. (2016) The cost of aircraft noise – Does it differ from road noise? A meta-analysis; in: *Journal of Air Transport Management*, Volume 57; pp. 138–142; doi:10.1016/j.jairtraman.2016.05.011

Krecl, P., Harrison, R.M., Johansson, C., Targino, A.C., Beddows, D.C., Ellermann, T., Lara, C., Ketzel, M. (2021) Long-term trends in nitrogen oxides concentrations and on-road vehicle emission factors in Copenhagen, London and Stockholm; in: *Environmental Pollution*, Volume 290; 118105; doi:10.1016/j.envpol.2021.118105

Lee, D.S., Fahey, D.W., Skowron, A., Allen, M.R., Burkhardt, U., Chen, Q., Doherty, S.J., Freeman, S., Forster, P.M., Fuglestvedt, J., Gettelman, A., De León, R.R., Lim, L.L., Lund, M.T., Millar, R.J., Owen, B., Penner, J.E., Pitari, G., Prather, M.J., Sausen, R., Wilcox, L.J. (2021) The contribution of global aviation to anthropogenic climate forcing for 2000 to 2018; in: *Atmospheric Environment*, Volume 244; 117834; doi:10.1016/j.atmosenv.2020.117834

Marais, K., Wolfe, P.J., Waitz, I.A. (2016) Air Transport and the Environment; in: Belobaba, P.P., Odoni, A.R., Barnhart, C. (Eds.): *The Global Airline Industry*; 2nd edition; John Wiley & Sons, Chichester

Meinshausen, M., Vogel, E., Nauels, A., Lorbacher, K., Meinshausen, N., Etheridge, D. M., Fraser, P.J., Montzka, S.A., Rayner, P.J., Trudinger, C.M., Krummel, P.B., Beyerle, U., Canadell, J.G., Daniel, J.S., Enting, I.G., Law, R.M., Lunder, C.R., O'Doherty, S., Prinn, R.G., Reimann, S., Rubino, M., Velders, G.J.M., Vollmer, M.K., Wang, R.H.J., Weiss, R. (2017) Historical greenhouse gas concentrations for climate modelling (CMIP6); in: *Geoscientific Model Development*, Volume 10; pp. 2057–2116; doi:10.5194/gmd-10-2057-2017

NASA – National Aeronautics and Space Administration (2025) GISS Surface Temperature Analysis; https://data.giss.nasa.gov/gistemp/

NOAA – National Oceanic and Atmospheric Administration (2025) Trends in Atmospheric Carbon Dioxide (CO2); https://gml.noaa.gov/ccgg/trends/gl_data.html

Sebastian, R.M., Louis, J. (2021) Understanding waste management at airports: A study on current practices and challenges based on literature review; in:

Renewable and Sustainable Energy Reviews, Volume 147; 111229; doi:10.1016/j.rser.2021.111229

Stansfeld, S.A., Berglund, B., Clark, C., Lopez-Barrio, I., Fischer, P., Öhrström, E., Haines, M.M., Head, J., Hygge, S., van Kamp, I., Berry, B.F., on behalf of the RANCH study team (2005) Aircraft and road traffic noise and children's cognition and health: a cross-national study; in: *Lancet*, Volume 365; pp. 942–949; doi:10.1016/s0140-6736(05)66660-3

Sulej, A.M., Polkowska, Z., Namieśnik, J. (2012) Pollutants in Airport Runoff Waters; in: *Critical Reviews in Environmental Science and Technology*, Volume 42 (16); pp. 1691–1734; doi:10.1080/10643389.2011.569873

Switzenbaum, M.S., Veltman, S., Mericas, D., Wagoner, B., Schoenberg, T. (2001) Best management practices for airport deicing stormwater; in: *Chemosphere*, Volume 43 (8); pp. 1051–1062; doi:10.1016/S0045-6535(00)00199-5

TRB – Transportation Research Board (2014) Critical Issues in Aviation and the Environment; *Transportation Research Circular* Number E-C184; https://onlinepubs.trb.org/onlinepubs/circulars/ec184.pdf

Teoh, R., Engberg, Z., Schumann, U., Voigt, C., Shapiro, M., Rohs, S., Stettler, M.E.J. (2024) Global aviation contrail climate effects from 2019 to 2021; in: *Atmospheric Chemistry and Physics*, Volume 24 (10); pp. 6071–6093; doi:10.5194/acp-24-6071-2024

Transport Canada (2021) Guidelines for Aircraft Ground – Icing Operations; https://tc.canada.ca/sites/default/files/2021-09/TP-14052E-6TH_EDITION.pdf

UBA – Umweltbundesamt (2020a) Ökologische Bewertung von Verkehrsarten; https://www.umweltbundesamt.de/sites/default/files/medien/479/publikationen/texte_156-2020_oekologische_bewertung_von_verkehrsarten_0.pdf; available in German only

UBA – Umweltbundesamt (2020b) Umweltfreundlich mobil! Ein ökologischer Verkehrsartenvergleich für den Personen- und Güterverkehr in Deutschland; https://www.umweltbundesamt.de/sites/default/files/medien/5750/publikationen/2021_fb_umweltfreundlich_mobil_bf.pdf; available in German only

Wang, W., Blanc-Betes, E., Khanna, M., Jiang, C., Guan, K., Guest, J.S., Field, J.L., DeLucia, E.H. (2025) Land conversion to energy crops for sustainable aviation fuel production reduces greenhouse gas emissions in the United States; in: *Communications Earth & Environment*, Volume 6; 963; doi:10.1038/s43247-025-02913-x

WHO – World Health Organization (2018) Environmental Noise Guidelines for the European Region; https://iris.who.int/bitstream/handle/10665/279952/9789289053563-eng.pdf

WHO – World Health Organization (2024) Ambient (outdoor) air pollution; https://www.who.int/news-room/fact-sheets/detail/ambient-(outdoor)-air-quality-and-health

Worldatlas (2025) The Largest Airports In The World By Size; https://www.worldatlas.com/transportation/the-largest-airports-in-the-world-by-size.html

Zhao, X., Taheripour, F., Malina, R., Staples, M.D., Tyner, W.E. (2021) Estimating induced land use change emissions for sustainable aviation biofuel pathways; in: *Science of the Total Environment*, Volume 779; 146238; doi:10.1016/j.scitotenv.2021.146238

5 Regulatory Framework and Political Initiatives – Environment

Introduction

The aviation industry is subject to a large number of multifaceted environmental regulations. In line with the global nature of aviation, many regulations are set on an international level, complemented by measures implemented by states and sometimes even regions. Against this background, this chapter first discusses the necessity of environmental policies from an economic perspective and provides a general overview of the different instruments available for policymakers. In a second step, it is shown how the major environmental problems caused by aviation, notably noise, pollution and its contribution to global warming, are addressed by different jurisdictions and industry bodies.

Some Principles of Environmental Economics

In a market economy, transactions are based on voluntary agreements. Therefore, it can be assumed that usually all partners in a market transaction benefit; otherwise, they would simply not agree to this transaction.[1] Moreover, the overall market outcome is characterised by an efficient allocation of scarce resources and an output structure that is in line with consumers' preferences.[2]

From an economic perspective, environmental problems are the result of specific market distortions (or market failures), so-called externalities.[3] In general, an externality is an effect of A's activity on B, which is not based on a mutual agreement. Such externalities can either be positive (external benefit) or negative (external cost), and A and B might be firms, households, or authorities, e.g., a state or a municipality. As an example, assume that person A is sitting in his backyard, trying to read a book, while A's neighbour B is mowing the lawn – or an aircraft operated by airline C is flying over A's house – and the sound emissions are limiting A's ability to concentrate on the book (i.e., they are perceived by A as noise).[4] In this simple example, there are only two entities, and the external effect is locally confined. However, the example of climate change shows that often many actors contribute to external effects and that these can also be long lasting or have global implications.

DOI: 10.4324/9781003608172-5

Market theory also states that prices are determined by the interaction of supply and demand. In the long term the (average) price that a producer receives for a product or a service has to cover at least the (average) costs of production. If not all costs of production have to be borne by the manufacturer, like in the case of negative externalities, from an overall perspective the market price of this product is too low, and consequently the output is too large. Therefore, externalities do not only lead to fairness (or distributional) problems – i.e., A suffers from B's activities – they also create overall economic inefficiencies that justify regulatory measures, i.e., environmental policy. Such policies are often referred to as an internalisation of external costs, e.g., via environmental taxes (see below).

It has to be pointed out that – in almost all cases – negative externalities are an unwanted by-product of some activity which is generating advantages (in economics often referred to as utility). For example, a cargo airline does not operate its aircraft during the night with the intention of disturbing somebody's sleep, but in order to ship products to customers that value an overnight delivery. Consequently, the level of externalities depends on the overall level of economic activity and the share of the different industries. For example, if the economy is growing and the average income of people is increasing, usually more people will travel, leading to an increase in emissions from the aviation industry (all else equal).

In order to reduce emissions, basically two options exist.[5] The first option is simply to reduce the level of activities. The second option is to change the emission intensity of the respective activity, in other words, to reduce the amount of emissions per output unit by technical or operational changes. However, as firms usually use the least-cost technology, these changes will entail additional cost (if the technology with lower emissions would be less costly, firms or households would use it anyway). Consequently, the (average) price of the product will rise, with the negative effects on demand depending on the respective price elasticity.

The marginal costs of emission reduction usually increase with the level of emission abatement. Assume that polluters are obliged to reduce emissions by 10%. Following the economic principle, they will choose the abatement option that satisfies this requirement at minimum cost (often referred to as low hanging fruits). If the polluters have to reduce emissions further, additional, more costly measures have to be taken.

Environmental Policy

Overview on Environmental Policy Instruments

Although sometimes emissions are reduced on a voluntary basis, e.g., households abstain from consuming goods that lead to negative externalities, in most cases governmental intervention is necessary to reduce emissions to a socially acceptable and/or economically efficient level. Therefore, it has to be decided which measures should be taken (i.e., the choice of instruments of environmental policy) and which entity should be in charge of the specific policy (i.e., the allocation of tasks and competencies between different jurisdictions).

There is a large variety of environmental policy instruments. One common classification distinguishes between "soft" measures, "command-and-control" instruments, economic (financial) incentives and emissions trading schemes.

- The government uses soft instruments in an attempt to persuade private entities to reduce emissions. Examples include the provision of reliable information about emissions (e.g., a comparison between specific emissions from rail and air travel) and a role model behaviour of the government. In general, these instruments are considered to be only partially effective, but they might also be used to support the implementation of other instruments.
- By command-and-control instruments, the government guides the behaviour of private individuals or firms. There are many different options for the use of such policies. The wording can be positive or negative, i.e., they may prescribe a specific behaviour or they may prohibit a specific behaviour, therefore implicitly allowing all other options. Some policies refer to products, and the strictest option would be a prohibition of products or services, e.g., the ban of certain domestic flights in France since 2020.[6] On the other hand, policies can be process related, i.e., the way in which a product or a service is produced. The latter is often named a technical regulation,[7] with emission standards for aircraft or ground vehicles as well-known examples.
- Economic (or financial) incentives do not prohibit or prescribe a certain behaviour but provide economic "rewards" and/or "penalties". The most obvious example is an emissions tax, i.e., a polluter has to pay the tax based on the amount of emissions. Especially in transportation, it is quite common to also include environmental incentive schemes into user charges (see the section on noise surcharges later in this chapter). Contrary to penalising emissions, the government may also reward an emissions reduction, especially by offering subsidies. Finally, the incentive may also be non-financial, for example scarce resources like slots at congested airports may be allocated based on criteria that are linked to the sustainability of the user.[8]
- The latest and probably most complex instrument for environmental protection is an emissions trading scheme (ETS). In order to emit one unit of a pollutant, a private entity has to own a specific allowance (sometimes also called permit or emissions right) which will become invalid after the entity has emitted one unit of the respective pollutant. These emissions rights are granted by the government in limited numbers. The initial allocation of the allowances may be done through auctioning or as a free allocation based on criteria such as past emissions

(grandfathering) or output metrics. After the initial allocation, permits can be traded between private entities,[9] leading to a market price for the allowances.

As mentioned above, different types of instruments can be combined, and they can be elements of policymakers' overall strategies, taking also into account different time horizons.

Pros and Cons of Different Instruments – A General Assessment

In economic studies, specific criteria are used to assess different instruments for environmental protection.[10] First of all, an instrument should achieve the environmental goal, often referred to as effectiveness or efficacy. Second, the environmental objective should be reached at minimum cost (static efficiency, assuming a given technology). Moreover, there should also be incentives to invest into research and development, with the aim of further reducing the costs of emissions reduction (dynamic efficiency). Finally, particularly with regard to the political dimension of environmental policy, the distributional effects of the instruments (i.e., the creation of "winners" and "losers") and side-effects on other objectives such as economic growth matter.

Although command-and-control instruments are quite widespread, they have several shortcomings. In most cases, technical standards refer to specific emissions based on some input or output unit, e.g., aircraft noise emissions per movement. However, the total emission level is determined by the number of emissions-causing activities and the average emission amount per activity unit. Therefore, even if standards are tightened, thereby reducing the average amount of emissions per activity unit, an increase in activities could lead to a constant or even rising overall emission level. This phenomenon can be observed in many transport markets, including aviation. Of course, it is also possible to limit the level of activities itself. In particular, at some airports the number of aircraft movements has been capped at a level which is below the technical capacity of the respective airport due to environmental reasons.[11]

From an economic perspective, the efficiency of command-and-control policies depends on several factors. As mentioned above, the government might prescribe the use of a certain technology, or it might cap the amount of (specific) emissions. If the regulation refers to the level of emissions, the polluter will use the cheapest option available to fulfil its obligation, thereby also minimising the individual costs of environmental protection. On the other hand, if the government determines which technology a firm has to use or which products might be offered, it may happen that policymakers will not choose the most efficient option, thereby creating avoidable cost of emissions reduction, i.e., inefficiencies. Similarly, if a regulation affects several entities, like in the case of carbon emissions reduction, it will be inefficient if it does not take different levels of abatement costs into account.[12] As the government usually is not informed about the individual levels of abatement costs – and polluters do not have an incentive to provide realistic information[13] – uniform regulations across different firms or industries usually will create an additional economic burden which increases with the degree of variance between abatement costs of the entities that are subject to the specific regulation. From a dynamic perspective, polluters must comply with the respective regulation, but they do not have strong incentives to go beyond this obligation and to invest into the development of new technologies (see Chapter 8 for a more detailed discussion of the business perspective on sustainability-related measures). However, they might expect a further tightening of a regulatory standard, providing some incentives to think about additional options for emissions reduction.

Like for command-and-control policies, there are also many different design options for economic incentives. In the following, the example of an emissions tax will be used, i.e., for each polluter the tax payment is calculated by multiplying a uniform tax rate with the respective number of emission units. The major advantage of such an emissions tax is its economic efficiency. Each polluter will compare its abatement costs with the tax rate and will reduce its emissions up to the point where the tax rate equals the marginal abatement costs of the last emission unit. From an overall perspective, this will lead to the avoidance of emissions at minimum cost. Moreover, as the remaining emissions are not free, each polluter has an incentive to investigate additional options for reducing emissions, i.e., to invest into research and development.

Assuming that the government does not have full information about the overall abatement costs, the environmental effects of an emissions tax are unknown to a certain extent. If the government targets a specific level of emissions, e.g., an emissions reduction by 10%, it might have to adjust the initial tax rate, sometimes referred to as a trial-and-error-process. Moreover, it will be necessary to adjust the tax rate, e.g., in line with the overall inflation rate as well as with economic or technological developments that lead to a shift of the marginal abatement cost function.

Although from an overall economic perspective, emission taxes are in general more efficient than a command-and-control policy, they often face stronger opposition from polluters. This is due to the fact that taxes have to be paid for all remaining emissions, whereas within a command-and-control scheme, the remáining emissions are essentially free of charge.

Other financial measures are usually less targeted than an emissions tax. As explained in more detail below, taxes on air transport services, e.g., a ticket tax, reduce output and thereby also emissions, but even a tax with a somehow differentiated tariff, e.g., a higher tax rate for long-haul flights, does not precisely capture the different amount of emissions caused by individual flights and passengers.

The most obvious advantage of an emissions trading scheme is its effectiveness. Emissions are capped at the level set by the government.[14] Moreover, as allowances are tradeable, leading to a uniform market price, each polluter will reduce its emissions as long as the marginal abatement costs are below the price of the allowances, resulting in an overall efficient emissions reduction. Polluters will also benefit from technological innovations as they would have to purchase fewer allowances or might sell allowances they no longer need (dynamic efficiency).

Compared to an emissions tax, the effects of an emission trading scheme are similar in many respects. In a hypothetical situation, in which the government is perfectly informed about marginal abatement costs in the economy, it could reach the same level of emissions reduction by introducing an environmental tax or an ETS. In this case, the price of an allowance would equal the respective tax rate. However, in a more realistic setting in which the overall marginal abatement cost function is unknown, the two instruments differ in terms of their inherent uncertainty. As mentioned above, the precise effect of an emissions tax on the amount of emissions is uncertain. On the other hand, if the government introduces an ETS, ex ante the price of the allowances is unknown (and it will most probably vary over time), creating some uncertainty for polluters.

Within an ETS, there are several design options. In particular, one option for allocating the allowances would be an auction. Like in the case of an emissions tax, this would lead to a monetary flow from the polluters to the state. As an alternative, the government might allocate some or even all allowances for free. Depending on the criteria used for this allocation (e.g., past emissions, output, number of employees, etc.), this will create winners and losers, i.e., some polluters will have to purchase allowances from other polluters. Consequently, the criteria for such free allocations are usually disputed. Moreover, within an ETS it might also be possible to allow a shift of emissions in time, i.e., using allowances issued in one year for emissions in following years (so called banking of allowances, e.g., in the EU ETS).

In reality, policymakers often use several instruments in parallel. For example, a technical standard might be combined with an emissions tax, creating additional incentives from a dynamic perspective. Especially with respect to emissions trading schemes it is often overlooked that additional measures usually do not change the total amount of emissions but only their regional and/or sectoral distribution (waterbed effect).[15]

Allocation of Competencies for Environmental Policy

In economics, also the allocation of competencies between different levels of government is analysed, discussing the pros and cons of centralisation vs. decentralisation.[16] In general, costs and benefits of political decisions should mostly stay within one jurisdiction, in other words, spatial externalities (spill-over effects) should be minimised. Moreover, it can be assumed that decision-makers on decentralised levels, e.g., a city, are better informed about citizens' preferences on this level than decision-makers on a more central level, e.g., the state. On the other hand, decentralised decision-making might lead to a higher level of transaction costs (e.g., if firms are providing services in jurisdictions with differing regulations that they have to adapt to) and the provision of public goods might be subject to economies of scale, leading to lower costs if the respective good is provided on a central level.

For example, noise is always a local problem. Therefore, one might argue that the decision on night curfews should basically be decentralised, giving each region the opportunity to balance between the costs and benefits of such a restriction. On the other hand, the effects of CO_2 emissions on the climate are global, calling also for a world-wide policy approach. Especially in the area of climate policies, there is always the danger of carbon leakages, i.e., measures taken by one jurisdiction that at least to some degree only shift emissions to other jurisdictions with less ambitious policies, maybe even increasing total emissions.

Aviation Subsidies and the Environment

In the public debate, subsidies granted to the aviation industry are often criticised.[17] In the narrow definition of national income accounting, subsidies are payments from the public sector to firms without the provision of a specific good or service in exchange. In a broader sense, a subsidy is any government measure that gives certain companies or industries an advantage over other companies or industries in similar circumstances. Against this background, subsidies include tax exemptions, but also exemptions from costly regulations, for example in the area of environmental protection. In the following, selected types of subsidies are discussed that are often characterised as environmentally harmful.[18]

In many jurisdictions, a general sales tax or value added tax is levied, usually as an ad valorem tax (i.e., x% of the net price of a product). International aviation is generally excluded from such taxes, and in some countries, domestic flights are also exempt from taxation, or a reduced tax rate applies.[19] This might not only distort intermodal competition, especially if domestic train services are subject to a higher tax rate, but it might also affect the overall structure of household consumption, favouring air travel over other types of leisure activities. At least for international flights, imposing a value added tax might cause some administrative difficulties. However, several countries levy passenger taxes (often referred to as ticket taxes – see box) which can be seen as an (imperfect) substitute for a general sales tax or value added tax levied also on air transport services.

Ticket Taxes in Aviation

Several countries, e.g., Australia, Germany and the UK, levy specific taxes on departing passengers, usually as a quantity tax.[20] The most simplistic design is a uniform tax per departing passenger. Other countries differentiate their ticket taxes depending on the traveller's final destination (usually with a higher tax rate for longer distances, but there are also some examples for higher tax rates for very short flights) and/or the class of travel (with higher tax rates for passengers travelling in business or first class). At hub airports, connecting passengers are usually exempted, and only very few countries also levy a tax on air cargo.

All other things being equal, ticket taxes have a negative impact on demand for air transport services and thus also reduce the environmental impact. However, these effects are not very specific, as ticket taxes do not differentiate between the technical standard of the aircraft, and the distance bands are usually rather broad. Leakage effects have been observed in border regions, with airlines and/or passengers switching to airports in neighbouring countries without air passenger taxes.

Taxes on fuel are common in many countries, usually levied as a quantity tax per unit of fuel, often with different tax rates for different types of fuel (e.g., gasoline vs. diesel). Commercial aviation[21] is usually exempted from fuel taxes, which is often criticised as a subsidy harmful to the environment. The legal framework is quite complex. The Chicago Convention (Art. 24) governs for international aviation that fuel on board an arriving aircraft must not be taxed once this fuel is used on the departing flight. Moreover, bilateral air service agreements usually exempt fuel purchased by the respective foreign airline from any kind of fuel taxation. As states do not want to discriminate their own airlines, this clause excludes almost all international flights from fuel taxation. Similarly, the European directive on energy taxation (2003/96/EC) allows fuel taxes for intra-European flights between two member states if both states bilaterally agree on such a tax. Finally, fuel taxation on domestic flights might be implemented by national law.[22]

The effect of this "fuel subsidy" depends on the type of flight as well as the tax rate. Assume, for example, that the average fuel burn per passenger on a short-haul flight (400 km) is 25 litres.[23] In this case, a 0.50 Euro tax per litre of fuel would lead to an additional average tax burden of 12.50 Euro per passenger. On long-haul flights, the additional tax burden might reach three-digit numbers.

Like in many other cases, one major reason for tax exemptions in the air transport industry are potential distortions of competition. If, for example, domestic flights were subject to a fuel tax, this would not only affect travellers between two cities within this country but also transfer passengers that travel via a domestic hub airport. Consequently, airlines offering connections via foreign hubs would benefit from a cost-advantage, even if the total fuel burn of such connections would be higher than on the connection via the domestic airport. Moreover, bilateral fuel taxes might also affect the tourism industry, if only levied on flights to destinations in one or few countries.

Subsidies might be granted to entities along the entire aviation value chain. However, one must distinguish between general subsidies that might be available to other industries as well (e.g., regional subsidies) and subsidies targeting the aviation industry. Moreover, subsidies may be granted in exceptional circumstances, e.g., during the COVID-19 pandemic.[24] The most obvious case for a distortion of competition are subsidies to domestic airlines, especially if these airlines are state-owned. In the European Union, rather strict limits exist for such state aid measures, which have even contributed to the market exit of airlines like Malév in Hungary.[25] In other jurisdictions, there are fewer restrictions on subsidies and usually transparency about such measures is limited. Moreover, infrastructure providers (airports and ANSPs) might receive subsidies, thereby lowering the operating costs for airlines. Whereas larger airports usually are profitable, many regional airports depend on subsidies. In the European Union, also state aid to regional airports is subject to some restrictions,[26] especially if it might distort competition between airlines. Finally, there are long and to some degree ongoing controversies on subsidies granted to aircraft and engines manufacturers.[27]

Environmental Policy in Practice – Aircraft Noise

ICAO's Balanced Approach

On the one hand, aircraft noise is a typical example for an environmental problem at the local level, justifying decentralised decision-making to take different local priorities and conditions into account. On the other hand, a "patchwork" of differentiated local rules is an obstacle for the development of aviation. Moreover, local rules might only provide limited incentives for aircraft manufacturers to invest into the development of less noisy aircraft. Against this background, ICAO has developed its balanced approach to aircraft noise management, which has become a guideline to policymakers in many jurisdictions.[28] As a prerequisite for assessing the magnitude of noise problems as well as the effects of noise mitigation measures, airports in many countries are obliged to monitor noise pollution in their surrounding areas.

The ICAO balanced approach consists of four elements, with a certain prioritisation for the first three elements:

- Land-Use Planning and Management,
- Noise Reduction at Source,
- Noise Abatement Operational Procedures,
- Operating Restrictions.

Land-Use Planning and Management

By limiting or even prohibiting new construction activities in the vicinity of airports, based on (projected) noise exposure levels, future conflicts might be avoided or at least reduced.[29] Land-use planning and other passive noise protection measures are typically governed by local or regional authorities. As a result, implementation varies by country and airport.

Typically, noise protection zones are established around commercial[30] airports to mitigate the impact of aircraft noise on surrounding communities. Within these zones, specific restrictions on new construction apply,[31] and landowners may be eligible for receiving financial support for noise insulation measures on existing buildings or financial compensations for a decline in property values. The applicable rules depend not only on the noise exposure level (sometimes with further differentiations between noise during daytime and night-time), but also on the intended use of the building. The strictest limitations apply to noise-sensitive uses such as schools and hospitals, followed by residential buildings, while industrial developments are generally less restricted. In addition to mandatory measures, airports might purchase land (or buildings) in areas subject to noise exposure on a voluntary basis, thereby being able to better steer the use of this land and reduce potential conflicts with residents and/or landowners.

Noise Reduction at Source

The probably most obvious option for an active mitigation of noise problems is the reduction of aircraft's sound emissions. The different technical options are explained in detail in Chapter 6. Already in 1969, Part 36 of the US Federal Aviation Act was passed, introducing maximum noise emissions for aircraft. On a global level, Annex 16, Volume 1, to the Chicago Convention became applicable in 1972. The limits specified in Annex 16 refer to the noise emissions of an aircraft at three measuring points (flyover, lateral and approach) and depend on the respective maximum take-off mass (MTOM).[32]

The ICAO standard defined in Annex 16, Volume 1, applies to the type certification of new aircraft models. In the following, only standards for noise emissions of subsonic jet aircraft are discussed.[33] The first version of the ICAO noise limits is Chapter 2 of Annex 16, Volume 1, from 1972. In 1977 these standards were tightened for the first time (Chapter 3). Chapter 4 regulates type certifications between 1 January 2006 and 31 December 2017, and the current version (Chapter 14) in general refers to type certifications after 31 December 2017.[34]

The limits defined in Annex 16, Chapter 2 and Chapter 3 are specific for the three measurement points mentioned above. Chapter 4 and Chapter 14 take a more flexible approach, prescribing that the sum of the respective Chapter 3 limits has to be undercut by a certain amount. Nevertheless, there are also some limitations to the use of trade-offs between emissions at the three measurement points.

Annex 16 only refers to the type certification of aircraft. Consequently, aircraft models with an existing type certification might still be produced, sold and operated. Given the long development and production cycles in aircraft manufacturing, it is obvious that strengthening the ICAO limits does not immediately lead to reductions in noise exposure at airports. In order to speed up the dispersion of aircraft complying with more recent standards, some states regulate the fleet structure of domestic airlines, prohibiting the purchase of older aircraft (non-addition rule) and/or force airlines to reduce the share of older aircraft in their fleets (phase out rules). In the USA as well as in the EU[35] such measures were applied in the 1990s to withdraw aircraft from service that only met the limits specified in Annex 16, Chapter 2. In both cases, operating restrictions for foreign airlines were also introduced (see below).

Noise Abatement Operational Procedures

At the airport level, specific operational procedures for aircraft are mandatory, referring to the horizontal and/or vertical flight path. A more detailed description of such procedures will be provided in Chapter 6 of this book. It is generally accepted that such noise abatement measures must not compromise safety. However, a conflict with economic objectives may arise if the requirements lead to a reduction in airport capacity. Moreover, there might be trade-offs with other environmental objectives if noise reducing operations lead to a higher fuel burn and therefore higher CO_2 emissions.

In general, noise management at airports involves balancing two seemingly opposing strategies: (1) concentrating flights over areas with low population density and (2) dispersing traffic to avoid repetitive burdens on the same communities. Where feasible, flight paths are concentrated over uninhabited areas or industrial zones to limit overall population exposure. However, in cases where overflight of residential areas is unavoidable, especially in the vicinity of large urban airports, authorities often adopt strategies to spatially and temporally disperse noise to increase fairness in exposure and reduce long-term health effects on part of the surrounding population. A typical strategy is the use of preferential runway systems, where airports assign specific alternate runways or flight paths depending on wind conditions, time of day and community noise sensitivity.[36]

Operating Restrictions

Operating restrictions at airports might take many different forms and they might be harmonised within a jurisdiction or might be tailor-made for a specific airport. One example for a general restriction is the operational ban for aircraft not complying with Annex 16, Chapter 3 limits in the EU (Council Directive 92/14/EEC) which came into effect in 2002.[37] Moreover, several airports also prohibit movements with aircraft not complying with specific noise standards during certain periods, especially during the night-time. While such restrictions could reduce noise problems at certain airports, from the airlines' point of view they lead to greater complexity in the flight planning process and less flexibility in daily operations.

At least in the European Union, night curfews are quite common,[38] usually based on regional regulations within a national framework. While at some airports no movements are permitted[39] (e.g., Frankfurt Airport, between 11:00 PM and 5:00 AM)[40], the number of movements is limited at other airports, sometimes combined with additional restrictions on the noise emissions of the aircraft operating. Moreover, at some airports certain trade-offs might be used, i.e., the number of movements can be increased if the noise emissions per movement are reduced.[41]

For some airports, the maximum total number of movements is restricted below the technical capacity of the airport,[42] aiming at a reduction of noise emissions (and also gaseous emissions). However, this measure is not very targeted, and it might even provide incentives for the use of larger (i.e., noisier) aircraft.

In general, especially night curfews are the only option for enabling an undisturbed sleep during night-time. However, compared to other instruments they also cause rather high costs for airlines, depending on the respective business model. For cargo flights, operations during night-time are often crucial, and especially integrators like DHL and UPS depend on such night flights for operating their hubs. In the passenger market, all airlines and especially leisure and low-cost carriers aim at a high aircraft utilisation, and many of their passengers are willing to depart at a very early hour or arrive late at night. For all airlines, the basic disadvantage of night curfews is the limited flexibility in case of operational disruptions. There are several examples for flights that were not permitted to depart if they arrived at the runway one minute too late, forcing the airline to provide hotel accommodation for the affected passengers. Moreover, arriving aircraft were diverted to another airport in case they have started the approach shortly after the beginning of the curfew.[43] The airline industry often calls for a more flexible approach, however, if airlines anticipate this, they might adapt their schedules, leading to more flights during the night-time and similar conflicts at later hours.

Financial Incentives for Using Less Noisy Aircraft

In addition to the measures described above, which are part of ICAO's balanced approach, financial incentives might be provided for the use of less noisy aircraft. There are several design options, and a large variety of examples can be found at different airports, especially in Europe. First, the financial incentive might be provided via the government or via the airport. There are only few examples for noise taxes levied by the state, and even in these cases the revenue is often earmarked and transferred to the airport operator (e.g., the French Tax on Air Traffic Noise).[44] In most cases, airports have included incentives within their respective charging schemes, either as a modulation of movement-related charges or as a specific surcharge for noise emissions.[45] Moreover, the design of the incentive might be based on the technological standard of the aircraft or on the actual noise emissions.

In Europe, already in the late 1970s some airports included financial incentives for less noisy aircraft into their charging schemes (e.g., Düsseldorf). The movement-related charge, which is based on the maximum take-off mass of an aircraft, was modulated.[46] Today, similar schemes can still be observed at some airports, usually referring to the difference of the noise emissions measured during type certification and the respective Chapter 3 limits.[47]

At other airports specific noise surcharges have been implemented. They refer to the standardised noise emissions measured during type certification (e.g., Sweden),[48] the average noise emissions of a specific aircraft type at the respective airport (e.g., German airports like Frankfurt and Munich),[49] or the actual emissions of each aircraft movement.[50] Moreover, there might be different classes or categories, comprising all aircraft with emissions between x and y dB, or the airport uses a continuous formula for calculating the noise surcharge (often non-linear with an increasing marginal surcharge for each decibel).

Using specific noise emissions measured at an airport rather than the standardised emissions determined during type certification considers that aircraft are used for different purposes. For example, during departure a narrowbody aircraft operating on a short distance and with a rather low load factor will be less noisy than the same aircraft flying with maximum payload to a destination close to its maximum range. Moreover, operational procedures might matter, an aspect covered especially by the movement specific noise surcharge at Berlin Airport.

The elements described above might be combined, for example Frankfurt Airport levies a noise surcharge based on average noise emissions per aircraft type combined with a discount based on the difference between the noise emissions determined within the type certification and the Annex 16 limits.[51] Moreover, most airports levy higher charges and/or surcharges during night-time and/or in the period before and after a night curfew.

If an airport implements financial incentives for using less noisy aircraft, the overall framework for setting airport charges matters. Airports are often considered to be natural monopolies, and they are usually subject to a specific regulation. Assuming, for example, a simple revenue cap, noise-related charges must basically be revenue neutral, i.e., creating winners and losers amongst airlines. In some cases, airports levy specific noise surcharges to finance additional expenses, for example caused by an obligation to finance noise mitigation measures mentioned above.

Discussing the pros and cons of the many variants of different design options of noise surcharges would go beyond the scope of this publication. Consequently, the analysis is limited to some basic effects. In general, for each incentive one would have to distinguish between short- and long-term effects. In the short term, the fleets of the respective airlines are basically given. Therefore, different noise-related incentives at different airports might lead to changes in network planning and therefore a redistribution of noise. In particular, London Heathrow Airport levies a rather high noise surcharge for an Airbus A320ceo when compared to an A320neo. Consequently, airlines that operate both types prefer an A320neo to fly into Heathrow. As airports often decide on these incentives on a decentralised level[52] the overall allocation of the least noisy aircraft within a network might not be optimal from an overall perspective. In the long run, airlines might have an incentive to accelerate fleet modernisation. However, airport charges are only a small part of an airline's investment decision.

Local Air Quality

Technical Standards

Similar to the ICAO technical standards for limiting noise emissions, Annex 16 Volume 2 contains limits for pollutant emissions since 1982, in particular regarding smoke, unburned hydrocarbons (HC), carbon monoxide (CO), nitrogen oxides (NO_x) and non-volatile particulate matter (PM). The amount of the respective emissions is limited for a standardised landing and take-off (LTO) cycle (see Chapter 6 of this book). In general, the permitted amount of emissions depends on engine and operating conditions, in particular thrust and pressure ratio. Therefore, the absolute amount of emissions increases for larger, more powerful engines. Amongst the pollutants covered by Annex 16, only the limits for NO_x emissions were strengthened several times since Annex 16 Volume 2 has been implemented.

Annex 16 Volume 2 standards refer to aircraft engines. With respect to NO_x emissions, not only the date of manufacture of the first series model is determined but also the actual date of manufacture of each individual engine. Therefore – unlike Annex 16 Volume 1 – the standards for pollutants also prohibit the production of engines that do not comply with current standards.

Operating Restrictions

In order to reduce emissions during the LTO cycle, reducing taxi times is one suitable option. However, it is in the own interest of airlines to minimise taxi times as well as fuel consumption during taxiing. Therefore, no additional regulation seems to be necessary. However, at many airports the use of auxiliary power units (APUs) is restricted, forcing airlines to use energy sources provided locally by the airport (see also Chapter 6).

Financial Incentives for Using Aircraft with Low Emissions

Whereas at the airport level, noise surcharges are quite common, only a few airports also apply surcharges for emissions of NO_x and other pollutants. Especially in Germany, this instrument is quite widespread, with a linear tariff, i.e., x Euro for each unit of emissions during a standardised LTO cycle. Compared to noise surcharges, the absolute amount of pollution-related charges is much lower. The pros and cons are basically the same as in the case of noise surcharges.

CO_2 Emissions and Climate Change

Overview

There are basically three options for an airline to either reduce or compensate CO_2 emissions caused by aircraft operations. First, the specific fuel consumption can be reduced by technical changes to the aircraft (see Chapter 6 for more details) and also more efficient operations, e.g., less holding patterns at airports. However, there might be a rebound effect, i.e., if flights become more efficient, the costs of travel decrease, potentially leading to a higher demand. In other words, technical progress and operational improvements reduce specific emissions per passenger-kilometre, but not necessarily total CO_2 emissions. Second, airlines might use sustainable fuel, i.e., fuel that causes less CO_2 emissions per unit of fuel burned because it was produced by using either biomass or green electricity (see Chapter 8). Third, CO_2 emissions caused by aircraft could be offset by reducing emissions elsewhere or removing them from the atmosphere. There are basically two instruments for such measures. In some jurisdictions, aircraft emissions are included into an emissions trading scheme (ETS – see general explanation above) and airlines have to purchase allowances from other sectors if they increase their CO_2 emissions. Moreover, some airlines compensate their emissions by either paying for emissions reduction not covered by an ETS or by removing CO_2 from the atmosphere, e.g., by planting trees. The major difference between an ETS and offsetting programs is the coverage of the scheme. An ETS is a closed scheme with a cap on total emissions within the scheme, whereas offsets are outside of an ETS and therefore it is more difficult to verify that the respective measures lead to an emissions reduction or removal.

As an industry initiative, IATA (2025) and its member airlines have committed to achieving net zero CO_2 emissions by 2050. This ambition is structured around four pillars: the use of Sustainable Aviation Fuel (SAF), the development of new aircraft and propulsion technologies, operational and infrastructure improvements and carbon offsetting and removals, covering all options mentioned above (see Chapter 6).

ICAO's Policies – Emission Standards for Aircraft

In 2017, ICAO introduced CO_2 emissions standards for aircraft. Annex 16, Volume 3, to the Chicago Convention contains limits to the specific CO_2 emissions of subsonic jet and turboprop aircraft. One major challenge for the definition of CO_2 standards for aircraft is the dependence of fuel burn (and thereby CO_2 emissions) on the actual weight and the respective flight phase. In Annex 16 Volume 3 a rather complex calculation of permissible CO_2 emissions is specified, considering the transport capability (payload and range) as well as different phases of the flight (beginning, middle and end of cruise phase). The standard is not only mandatory for newly designed aircraft types but also for established aircraft models. Aircraft that do not meet the standard may no longer be produced after 2028.

As fuel burn is an important contributor to airlines' operating costs, it is debatable whether an additional standard for maximum fuel burn is necessary. In general, the standard would only be effective if it prevented the production of less fuel-efficient aircraft that would have been purchased by airlines without this standard.[53]

ICAO's Policies – CORSIA

In order to achieve Carbon Neutral Growth (CNG), ICAO in 2016 implemented the Carbon Offsetting and Reduction Scheme for International Aviation (CORSIA), coming into effect in 2019. This scheme, governed through Annex 16, Volume 4, to the Chicago Convention consists of several measures and phases. The general idea is to offset all CO_2 emissions beyond the emissions baseline (85% of emissions in 2019).[54] As a general limitation it has to be pointed out that, in line with ICAO's overall mandate, CORSIA only covers emissions on international flights. Domestic flights exclusively remain in the competency of the respective countries.

Within CORSIA, all airlines above a certain threshold (10,000 tonnes of CO_2 emissions per year) have to monitor, report and verify (MRV) their emissions on international flights (see also Chapter 7). Compensation requirements differ between several phases. In the first two phases (from 2021 to 2023 and from 2024 to 2026), participation in CORSIA is voluntary, and only flights between participating states are subject to offsetting obligations. Although the number of participating countries increased from 88 in the year 2021 to 130 in 2026,[55] several large aviation nations are still missing, especially Brazil, China, India and Russia. In 2025, the share of international aviation emissions covered by CORSIA is around 60%.[56] From 2027 onwards, participation is mandatory if the airlines registered in the respective country had a share of at least 0.5% of global CO_2 emissions on international flights in 2018. Some least developed and landlocked countries are exempted. If the inclusion of all countries responsible for more than 0.5% of aviation's global CO_2 emissions on international flights does not cover at least 90% of the respective global emissions, more countries would be included, based on their respective share in emissions. Due to the definition of the baseline and the slow recovery of the air transport industry after the end of the COVID-19 pandemic, 2024 was the first year in which airlines actually had to offset emissions within the CORSIA scheme.

ICAO has defined specific criteria for offsetting projects in order to be eligible within CORSIA (environmental integrity). However, some critics claim that these criteria are not strict enough, thereby often not leading to additional CO_2 reductions or removals.[57] Moreover, if airlines use SAF or lower carbon aviation fuels, as defined in Annex 16, Volume 4, the offsetting requirements are reduced.

From an airline's perspective, the obligation to purchase offsets leads to an additional financial burden, and the assignment of this obligation has to some degree been controversial. Industrialised countries, e.g., in North America and Europe, are characterised by high emissions per capita and rather low aviation growth rates. On the other hand, in countries like China and India, current emissions per capita are comparably low, but growth forecasts are above average. Consequently, an obligation to offset additional emissions at the airline or country level would have led to rather high additional burdens in these countries, whereas airlines from industrialised countries would have had to offset only few emissions in spite of their much higher baseline. Therefore, in the first years of CORSIA all airlines have to offset emissions based on the global emissions growth rate. For example, if this global growth rate would be 5%, all airlines covered by CORSIA would have to offset 5% of their emissions on routes subject to CORSIA, independent of their individual growth rate. In later years, beginning in 2033, the obligation to compensate will be based on global as well as airline specific emissions, with the respective shares determined by ICAO at a later stage. Consequently, the effect of CORSIA on airlines is quite similar to an increase in the fuel price, however limited to fuel burned on routes covered by CORSIA. In later years, the share of fuel for which offsets will have to be purchased will differ between airlines, thereby also affecting the competitive situation.

Emissions Trading Schemes in the EU and Elsewhere

In 2011, the European Union introduced an emissions trading scheme for so-called large installations, basically referring to the manufacturing sector, e.g., refineries, cookeries and steel production (Directive 2003/87/EC). As households and small- and medium-sized enterprises are not yet included, the ETS covers only a part of total CO_2 emissions in the EU. From 2012, the emissions of the air transport industry were also included into the ETS.[58] Originally, the EU wanted to include all flights arriving at or departing from an EU airport. However, this led to strong opposition from several non-EU countries, in particular the USA and China.[59] Therefore, the EU decided that only intra-EU flights (including domestic flights) are covered by the ETS.[60]

As mentioned above, there are several ways of allocating allowances within an ETS. In the first years after the introduction of the ETS, the largest share of allowances was given to airlines for free, based on transport services performed in a base year, specified for each ETS reference period. However, the total amount of allowances provided to the air transport industry was fixed. Consequently, due to the growth of the air transport market, airlines had to purchase an increasing number of allowances from other sectors.[61] More recently, the EU reduced the share of free allowances, and from 2026 onwards, basically all aviation allowances shall be auctioned.

The price of allowances is determined by supply and demand. Whereas in the first years after the introduction of the ETS prices were quite low, since the

2020s the market is characterised by higher prices (in 2024, the spot market price for allowances was on average 65 Euro/tonne).[62] On intra-European origin-and-destination markets, the scheme does not cause intra-modal distortions, and the higher competitiveness of inter-modal alternatives like high-speed rail is an intended effect of the ETS. However, on some markets the ETS has similar competition effects as an intra-European fuel tax.[63] Airlines from non-EU countries that offer transfers outside of the EU, e.g., in Türkiye or the United Arab Emirates, are not subject to the ETS and therefore possess a competitive advantage compared to European hub airlines. This is of particular relevance for traffic flows to Asia due to the geographic location of these extra-EU hubs.

As the EU ETS also covers international flights between EU countries, these flights are exempted from CORSIA to avoid a double burden. Moreover, from 2027 onwards, the EU wants to include international flights to countries not participating in CORSIA in the ETS.

The additional costs incurred by airlines as a result of the ETS are flight specific. In the example of an intra-European flight with a narrow-body aircraft carrying 150 passengers over a distance of around 1,500 km and consuming 7,000 kg of fuel, around 22 tonnes of CO_2 are emitted. If the airline has to purchase the necessary allowances at a price of EUR 100, this will result in additional costs of EUR 2,200, or around EUR 15 per passenger.

In 2021, the People's Republic of China introduced a nationwide ETS. However, this ETS has a smaller (but growing) coverage in terms of industry sectors. Moreover, it is not a cap-and-trade scheme as outlined above, but a baseline-and-credit system which does not limit the absolute amount of emissions.[64] Aviation has not been included yet.[65]

Promoting SAF – Subsidies and Mandates

For the promotion of Sustainable Aviation Fuel basically two options exist. Option one is to provide subsidies to SAF producers, thereby reducing or even eliminating the cost differential compared to fossil fuels.[66] This strategy is used especially in the USA with the Inflation Reduction Act providing tax credits for SAF production.[67]

Alternatively, in some jurisdictions, airlines are legally obliged to use a certain proportion of SAF on their flights. This policy has been chosen by the European Union,[68] so-called SAF mandates.[69] From 2025 onwards, fuel suppliers have to provide fuel with a certain share of SAF at EU airports, and airlines have to prove that they have actually purchased the respective share of SAF. The latter was implemented by the EU in order to prevent so-called tankering. Tankering means to fuel the aircraft at airports where fuel is cheaper. However, due to technical constraints (limits to the amount of fuel that can be taken on board as well as to the maximum take-off and/or landing weight) and economic considerations (additional fuel burn due to the higher weight), this strategy is only relevant on few, rather short markets to airports outside of the EU. The SAF share increases from 2% in 2025 stepwise to 70% in 2050, complemented with a subquota for synthetic SAF (often named eSAF), starting at 1.2% in 2030 and rising up to 35% in 2050, i.e., half of the SAF will have to be eSAF in 2050 (for these different types of SAF see Chapter 8). The additional financial burden for the air transport industry depends on the share of SAF that has to be used and the price differential between SAF and fossil fuel.[70]

Unlike the ETS, the SAF mandate not only covers all flights within the EU but also all departing flights from the EU to non-EU destinations. However, there are again competitive distortions on transfer flights. If the connection to a non-EU airport is offered via a hub within the EU, both segments are subject to the SAF mandate. On the other hand, for transfers at a non-EU hub, only the first segment is subject to a SAF mandate.

Regulating Non-CO_2 Effects

So far, there are no regulations in place that directly aim at a reduction of non-CO_2 effects from air transport. However, the EU implemented a reporting obligation for non-CO_2 effects which came into effect in 2025 (see also Chapter 7):[71] "By the end of 2027, the Commission will deliver a report on the results and if appropriate, will make a legislative proposal to address non-CO_2 effects of aviation."[72]

With respect to contrail formation, states might introduce command-and-control measures, i.e., blocking certain parts of the airspace in which contrail formation would be expected. Alternatively, the use of such airspaces might be subject to higher en-route charges (financial incentives), or – if an ETS is already in place – there might be an obligation to submit additional allowances if contrails are formed during a flight.

Key Takeaways

- From an economic perspective, environmental problems can be characterised as market failure, justifying dedicated policy measures.
- Policymakers can use "soft" instruments, command-and-control measures, financial incentives and emissions trading schemes (ETS). Depending on the environmental problem which shall be tackled, these instruments differ with respect to their efficacy and efficiency.
- When implementing environmental policy instruments, side effects on other economic and political objectives matter. Furthermore, it has to be decided which governmental level should be granted the competency for imposing measures.
- Political measures against aircraft noise have the longest tradition in aviation environmental policy. ICAO has introduced noise standards for the type certification of aircraft which have been strengthened several

times (Annex 16, Volume 1). States have imposed restrictions for domestic airlines' fleets as well as general operating restrictions at airports.

- On the local level, especially European airports have included incentives for less noisy aircraft into their charging schemes. Local and regional governments have introduced night curfews as well as restrictions on land use in the vicinity of airports (passive noise protection).
- Local air quality has received much less attention from policymakers than aircraft noise. Nevertheless, ICAO has introduced technical standards for aircraft engines (Annex 16, Volume 2) and some airports offer financial incentives for using aircraft with relatively low emissions.
- In the last 10 to 15 years, an increasing number of measures have been introduced which aim at reducing CO_2 emissions from aviation. ICAO's CORSIA is the first example for a scheme that globally aims at carbon neutral growth. However, only emissions above a baseline are compensated and the environmental integrity of the programme is to some degree disputed.
- The EU ETS, covering only Intra-EU flights, is more ambitious, but also leads to some competitive distortions especially affecting European hub carriers that compete with airlines that operate hubs outside of but rather close to the EU.
- Like some other countries the EU has implemented a SAF mandate, further increasing airlines' costs, whereas other countries, in particular the USA, rely more on subsidies for SAF producers.
- Currently, measures to also tackle the non-CO_2 effects of air transport are discussed more intensely, but without specific measures introduced so far.
- Finally, with respect to the overall economic framework, tax exemptions and other subsidies granted to the air transport industry have been discussed more critically, and several states introduced specific ticket taxes or similar instruments in order to compensate other privileges that the air transport industry historically enjoys.

Notes

1 There might be exceptions, e.g., if one participant exerts pressure on transaction partners or is using some kind of fraud (immoral contracts).
2 Furthermore, a competitive market process results in an income distribution that is based on market performance, and the market is providing incentives for innovations as well as quick adaptations to changes in the general framework.
3 The concept of externalities goes back to Pigou (1920). A fundamental analysis of external costs and different instruments of environmental policy can be found in any textbook on environmental economics, e.g., Buchholz, Reif and Rübbelke (2024).
4 There are also other types of externalities that are not directly related to environmental problems, e.g., congestion.
5 A third option is to reduce the damage of the emissions, especially passive noise protection.
6 See Txapartegi, Cazcarro and Galarraga (2024). However, in this paper the authors ignore the waterbed effect, i.e., the shifting of emissions within an ETS (see below), note 15.

7 In contrast, economic regulation determines market results like the number of suppliers, output quantity and/or prices.
8 In several jurisdictions, the right to operate at an airport at a specific point in time (slot) is allocated based on the historic use of the respective slot (so-called grandfather rights), often complemented by additional criteria in favour of new market entrants. It has been suggested to also consider the noise level of the aircraft for slot allocation, providing an economic incentive for airlines to operate with less noisy aircraft, see for example Feng et al. (2023).
9 Although the term emissions trading scheme is common for this instrument, it should be clarified that entities do not trade emissions but emissions rights (allowances).
10 For an analysis of instruments to reduce aviation's climate impact see Fichert, Forsyth and Niemeier (2020).
11 See for example AMS Schiphol (2024).
12 For example, as the air transport industry is a hard to abate sector, it would be inefficient to limit the overall share of aviation's emission in the economy.
13 If polluters expect that the government will prescribe higher levels of emissions reduction to polluters with lower abatement costs, polluters have an incentive to overstate their costs of emissions reduction in the public debate.
14 Of course, there is a general risk of fraud, i.e., polluters emitting without possessing the respective allowances. However, the compliance with governmental regulations must be monitored for all other instruments as well.
15 Like emissions within an ETS, the total amount of water in a waterbed is given, and putting pressure on one side of the waterbed will only change the quantity of water in different parts of the bed, but not the overall quantity. In theory, it is possible that the other instruments are so strict that the level of emissions is below the maximum set by the number of allowances. In such a situation, the ETS would be useless (and the price of the allowances would drop to zero).
16 This area of research is often referred to as fiscal federalism, see especially Oates (1972), or the economics of multi-level government.
17 See for example Klarwein (2023). An overview on different types of subsidies in aviation is provided by Gössling, Fichert and Forsyth (2017).
18 See for example Burger and Bretschneider (2022).
19 For an overview on taxation of air transport with a specific focus on the EU see CE Delft (2019).
20 On ticket taxes see for example Ricardo Energy & Environment (2021).
21 Within the general aviation segment, fuel taxes are levied in some countries for private flights.
22 In the Netherlands, a fuel tax was levied for domestic flights between 2005 and 2011 (see Posada, Culbert and Wendenburg 2025). However, there are only very few domestic flights in the Netherlands.
23 The average fuel burn per passenger depends in particular on the total fuel burn of the aircraft on this specific route, on the number of seats, as well as on the load factor.
24 An overview on subsidies granted to European airlines is provided by Martín-Domingo and Martin (2022).
25 EC (2012).
26 EC (2014).
27 For example Truxal (2024).
28 See ICAO (2008) – first edition from 2004 – as well as ICAO Assembly resolutions A33–7 from 2001 and A39–1 from 2016. Regulators around the world refer to this framework, e.g., EU Regulation 598/2014.
29 If additional residential areas are developed in the vicinity of airports, this is sometimes referred to as encroachment.
30 In some countries, noise protections zones also exist for military airports.

31 In some cases, planning authorities may even consider the removal or relocation of existing infrastructure to reduce long-term exposure in high-noise areas.
32 For noise emissions during departure (flyover), the number of engines matters additionally.
33 For example, Chapters 5, 6 and 10 deal with propeller driven aircraft, Chapter 12 refers to supersonic aeroplanes.
34 There are specific rules for aircraft with a relatively low MTOM.
35 Council Directive 92/14/EEC.
36 For example, London Heathrow (2025) implemented a structured runway alternation programme to provide predictable noise relief to communities living under approach and departure paths. Under normal daytime operations with westerly winds, arriving aircraft use one runway until 3:00 PM and then switch to the other for the rest of the day. This scheduled alternation helps distribute noise exposure more evenly between neighbourhoods.
37 The directive included exemptions for aircraft registered in selected developing countries.
38 The definition of the curfew period differs between airports. Moreover, at some airports operating bans during the night are complemented by a reduction of the number of movements in the hours before and after the curfew period (e.g., Frankfurt).
39 Of course, there are exceptions, especially in case of emergencies but also if operations during the day were subject to extraordinary distortions (e.g., weather related).
40 Fraport (2025).
41 One example is the quota count system implemented at several airports in the UK. See UK CAA (2025), Annex A.
42 See Niemeier (2013) for the example of Düsseldorf Airport, Germany.
43 Usually, there is some flexibility for delayed flights, but with a clear threshold.
44 Ministères Transition Écologique et al. (2025).
45 According to the German Air transport act it is even mandatory to include noise-related elements into the charging scheme.
46 Aircraft certified according to Annex 16 Chapter 3 paid a lower rate per ton of MTOM than aircraft not complying with the Chapter 3 limits.
47 One example is Amsterdam Airport with seven noise categories ranging from less than 11 EPNdB difference to more than 27 EPNdB difference (Schiphol Airport 2025).
48 Swedavia Airports (2025).
49 Frankfurt Airport (2025) and Munich Airport (2025).
50 This is a rather new approach implemented for the first time in Berlin in the early 2020s. See Flughafen Berlin Brandenburg GmbH (2025).
51 Frankfurt Airport (2025).
52 In some states the charges are set at a national level with different noise incentives, e.g., Sweden, see Swedavia Airports (2025).
53 In 2025, Boeing already asked for a waiver in order to continue producing 777Fs after 2027. See Jeffrey (2025).
54 Originally, when introducing CORSIA it was decided that the average emissions of the years 2019 and 2020 should constitute the emissions baseline. However, due to the severe effect of the COVID-19 pandemic on the air transport industry and thereby also on its emissions, there was a controversial debate on adapting the baseline. Finally, it was decided that between 2024 and 2035, 85% of the emissions in 2019 should serve as the emissions baseline.
55 For a list of participating countries, see ICAO (2025).
56 Aviation Benefits Beyond Borders (2025).
57 T&E (2021).
58 EC Directive 2008/101/EC.
59 Lewis and Volcovici (2012).

60 There are some exemptions, e.g., for smaller carriers. In general, based on separate agreements the EU ETS also covers Iceland, Liechtenstein and Norway.

61 Technically, the EU differentiates between allowances and aviation allowances. Whereas allowances can be used by any firm, aviation allowances can only be used by airlines. This prevents the airlines from selling allowances to other sectors. However, except for the period affected by the COVID-19 pandemic, the air transport industry always had to purchase allowances from other sectors, making the differentiation between allowances and aviation allowances irrelevant.

62 ESMA (2025).

63 For an analysis of the competitive effects of the ETS in the aviation market see Peter et al. (2016).

64 For a concise comparison of the ETS in both countries see GIZ (2025).

65 Although the aviation sector will not be included in the near term, several regional ETS pilots have begun experimenting with aviation inclusion. For example, Guangdong's allocation plan includes aviation enterprises, categorised by transport type and service model. Full-service carriers receive allowances via a benchmark method based on aircraft type and transport turnover, while other carriers use historical intensity metrics.

66 Increasing costs for using fossil fuel like a fuel tax and an ETS also improve the competitiveness of SAFs but lead to an overall cost increase for the air transport industry.

67 See IRS (2025). Changes to the scheme were made as part of the introduction of the One Big Beautiful Bill.

68 The EU is banking on a mix of quotas (ReFuelEU) and production subsidies (e.g., the Sustainable Innovation Fund) to scale up SAF.

69 Other countries with SAF mandates are for example Brazil, India and Japan. See T&E (2024).

70 If fuel suppliers cannot fulfil their obligations, they will have to pay penalties. Airlines complain that these additional costs are also passed on onto fuel users, see IATA (2025).

71 Commission Implementing Regulation (EU) 2024/2493

72 EC (2025).

References

AMS Schiphol (2024) Government commits to fewer flights at Schiphol; https://www.schiphol.nl/en/you-and-schiphol/news/government-commits-to-fewer-flights-at-schiphol/

Aviation Benefits Beyond Borders (2025) CORSIA explained; https://aviationbenefits.org/other-environmental-challenges/climate-action/market-based-measures/corsia/corsia-explained/

Buchholz, W., Reif, C., Rübbelke, D. (2024) *Foundations of Environmental Economics*, 2nd edition; Springer, Cham

Burger, A., Bretschneider, W. (2022) Environmentally Harmful Subsidies in Germany; Umweltbundesamt Texte 117/2022; Dessau-Roßlau; https://www.umweltbundesamt.de/sites/default/files/medien/479/publikationen/texte_117-2022_environmentally-harmful-subsidies.pdf

CE Delft (2019) Taxes in the field of aviation and their impact, Final report; Commissioned by the European Commission, DG MOVE; Luxembourg; https://cedelft.eu/wp-content/uploads/sites/2/2021/03/CE_Delft_7M16_taxes_in_the_field_of_aviation_and_their_impact.pdf

EC – European Commission (2012) State aid: Commission orders Hungary to recover incompatible state aid from national air carrier Malév; Press Release, 9 January; https://ec.europa.eu/commission/presscorner/detail/en/ip_12_7

EC – European Commission (2014) Communication from the Commission, Guidelines on State aid to airports and airlines; (2014/C 99/03); OJ C99/3; https://eur-lex.europa.eu/legal-content/EN/TXT/?uri=CELEX:52014XC0404%2801%29

EC – European Commission (2025) Reducing emissions from aviation; https://climate.ec.europa.eu/eu-action/transport-decarbonisation/reducing-emissions-aviation_en

ESMA – European Securities and Markets Authority (2025) EU carbon markets 2025; Paris; https://www.esma.europa.eu/sites/default/files/2025-10/ESMA50-481369926-30552_Carbon_Markets_Report_2025.pdf

Feng, H., Hu, R., Wang, D., Zhang, J., Wu, C. (2023): Bi-objective airport slot scheduling considering scheduling efficiency and noise abatement; in: *Transportation Research Part D: Transport and Environment*, Volume 115; 103591; doi:10.1016/j.trd.2022.103591

Fichert, F., Forsyth, P., Niemeier, H.-M. (2020) Air transport and the challenge of climate change – How aviation climate change policies work; in: Fichert, F., Forsyth, P., Niemeier, H.-M. (Eds.) *Aviation and Climate Change. Economic Perspectives on Greenhouse Gas Reduction Policies*; Routledge, Abingdon / New York; pp. 47–78

Flughafen Berlin Brandenburg GmbH (2025) Charges Regulation Berlin Brandenburg Airport; Berlin; https://corporate.berlin-airport.de/content/dam/corporate/no-language/geschaeftspartner/entgelte/entgeltordnung-ber.pdf

Frankfurt Airport (2025) Airport Charges According to Art. 19b Air Traffic Act (LuftVG); Frankfurt; https://www.fraport.com/content/dam/fraport-company/documents/geschaeftsfelder/betrieb/flughafenentgelte/nutzungsbedingungen-und-entgelte/Entgelte%202026_Final_Onlineversion.pdf/_jcr_content/renditions/original.media_file.download_attachment.file/Entgelte%202026_Final_Onlineversion.pdf

Fraport (2025) Operating hours: From 5:00 A.M. to 11.00 P.M. CEST; https://www.fraport.com/en/business-areas/operations/airside-and-terminal-operations/hours-of-operation.html

GIZ – Deutsche Gesellschaft für Internationale Zusammenarbeit (2025) China and the EU – A comparison of two emissions trading systems; https://www.dena.de/fileadmin/dena/Publikationen/PDFs/2025/ENTRANS/dena_FS_Entrans_ETS_China_and_the_EU-A_Comparison_Web-PDF.pdf

Gössling, S., Fichert, F., Forsyth, P. (2017) Subsidies in Aviation; in: *Sustainability*, Volume 9 (8); pp. 1295–1314; doi:10.3390/su9081295

IATA – International Air Transport Association (2025) Progressing Towards Net Zero Carbon Emissions by 2050; https://www.iata.org/en/iata-repository/pressroom/presentations/progressing-towards-net-zero-carbon-emissions-by-2050/

ICAO – International Civil Aviation Organization (2008) *Guidance on the Balanced Approach to Aircraft Noise Management*; Doc. 9829, 2nd edition; Montreal; https://www.icao.int/sites/default/files/sp-files/environmental-protection/Documents/Publications/Guidance_BalancedApproach_Noise.pdf

ICAO – International Civil Aviation Organization (2025) CORSIA States for Chapter 3 State Pairs; https://www.icao.int/CORSIA/corsia-states-chapter-3-state-pairs

IRS – Internal Revenue Service (2025) Sustainable aviation fuel credit; https://www.irs.gov/credits-deductions/businesses/sustainable-aviation-fuel-credit

Jeffrey, R. (2025) Boeing seeks emissions exemption to sell 777Fs beyond 2027; Air Cargo News; https://www.aircargonews.net/airlines/boeing-seeks-emissions-exemption-to-sell-777fs-beyond-2027/1081050.article
Klarwein, S. (2023) Why It Is Time to End Aviation Subsidies; https://earth.org/aviation-subsidies/
Lewis, B., Volcovici, V. (2012) U.S., China turned EU powers against airline pollution law; Reuters; https://www.reuters.com/article/us-eu-airlines-climate/insight-u-s-china-turnedeu-powers-against-airline-pollution-law-idUSBRE8B801H20121210/
Martín-Domingo, L., Martin, J.C. (2022); The Effect of COVID-Related EU State Aid on the Level Playing Field for Airlines; in: *Sustainability*, Volume 14 (4); 2368; doi:10.3390/su14042368
Ministères Transition Écologique et al. (2025) Aeronautical taxes; https://www.ecologie.gouv.fr/en/public-policies/aeronautical-taxes
Munich Airport (2025) Tariff Regulations, Part 1; Munich; https://www.munich-airport.com/_b/0000000000000029425085bb6759849e/Munich-Airport-Tariff-Regulations-2025.pdf
Niemeier, H.-M. (2013) Expanding Airport Capacity under Constraints in Large Urban Areas: The German Experience; OECD International Transport Forum, Discussion Paper 2013/04; Paris; https://www.oecd.org/content/dam/oecd/en/publications/reports/2013/03/expanding-airport-capacity-under-constraints-in-large-urban-areas_g17a22b9/5k46n45fgtvc-en.pdf
Oates, W.E. (1972) *Fiscal Federalism*; Harcourt Brace Jovanovich, New York, Chicago, San Francisco, Atlanta
Peter, M., Bertschmann, D., Faber, J., Smit, M., van Velzen, A. (2016) The EU Emission Trading Schemes' effects on the competitive situation within national and international aviation; Commissioned by the German Federal Environment Agency; Dessau-Roßlau; https://www.umweltbundesamt.de/sites/default/files/medien/378/publikationen/climate_change_22_2016_the_eu_emission_trading_schemes_effects.pdf
Pigou, A.C. (1920) *The Economics of Welfare*; Macmillan, London
Posada, E., Culbert, V., Wendenburg, A. (2025) Fossil Fuel Subsidy Reform in Aviation and Shipping; International Institute for Sustainable Development (IISD); Winnipeg; https://www.iisd.org/system/files/2025-02/fossil-fuel-subsidy-reform-aviation-shipping.pdf
Ricardo Energy & Environment (2021) *Study on the taxation of the air transport sector*. Final report. Commissioned by the European Commission DGTAXUD; London; https://taxation-customs.ec.europa.eu/system/files/2021-07/Aviation-Taxation-Report.pdf
Schiphol Airport (2025) *Charges and Conditions April 1, 2025*; Amsterdam; https://assets.ctfassets.net/biom0eqyyi6b/3liHCYk4gIQ3nGgkQdWeDQ/70791a8c24532acf3efd12fb2feea5dd/Setting_Charges__Conditions_1_April_2025_ACM_decision.pdf
Swedavia Airports (2025) Airport Charges & Conditions of Services; https://www.swedavia.com/contentassets/9514f15d5b9440feb72b4ff3cb3854d8/swedavia-airport-charges-and-conditions-of-services-2025-v.-1.1.pdf
T&E – Transport and Environment (2024) The SAF Observatory; https://uploads.transportenvironment.org/production/files/202410_SAF-Observatory_Briefing-2.pdf
T&E – Transport and Environment (2021) Corsia: worst option for the climate; https://uploads.transportenvironment.org/production/files/2021_03_Briefing_Corsia_EU_assessement_2021.pdf

Truxal, S. (2024) State subsidies and aircraft financing in the EU, USA, and China: a balancing act; in: *United Law Review*, Volume 29 (1); pp. 17–29; doi:10.1093/ulr/unae026

Txapartegi, A., Cazcarro, I., Galarraga, I. (2024) Short-haul flights ban in France: Relevant potential but yet modest effects on GHG emissions reduction; in: *Ecological Economics*, Volume 224; 108289; doi:10.1016/j.ecolecon.2024.108289

UK CAA – Civil Aviation Authority (2025) Aircraft movements, fleet mix and noise quota usage at Gatwick, Heathrow and Stansted, CAP 3013; London; https://www.caa.co.uk/publication/download/22569

6 Measures Taken by the Aviation Industry – Environment

Introduction

In Chapter 4, aircraft noise, pollution and global warming – the most important environmental problems associated with aviation – have been presented, complemented by additional topics such as land use and waste generation. Chapter 5 provided an overview on the regulatory framework, that in some cases also prescribes specific technologies or procedures.

This chapter explains technical as well as operational options for reducing emissions and environmental damage, again based on the respective type of emissions. However, several measures might affect more than one environmental problem. For example, modern aircraft are more fuel efficient than older models – thereby reducing CO_2 emissions – and they are usually less noisy. Therefore, a fleet modernisation might benefit the environment on a global as well as on a local level. However, there might also be trade-offs in other areas, e.g., if fuel-efficient aircraft emit more pollutants like NO_x.[1] If there is no technological solution to such trade-offs, decision-makers have to weigh the respective effects in order to decide on the preferred technological pathway. Moreover, it has to be pointed out that many measures require joint efforts by at least two stakeholders, a topic that will be discussed in more detail in Chapters 8 and 10, respectively.

Noise Reduction

As already mentioned in Chapter 5, active noise abatement reduces sound emissions or changes their spatial distribution whereas passive noise reduction is based on land use planning (limiting the use of noise-affected areas for residential and other noise-sensitive purposes) and the insulation of buildings, thereby reducing noise levels within. As shown in Chapter 5, passive noise protection is basically governed by the state and/or municipalities, often pledging airports to finance for example noise insulation programs. Moreover, airports might also purchase land in noise-affected areas or provide additional financial assistance for sound-proofing buildings on a voluntary basis. For example, Fraport AG, the operator of Frankfurt Airport, implemented a

DOI: 10.4324/9781003608172-6

voluntary purchase programme after the opening of its fourth runway, offering to buy residential properties affected by low overflight altitudes at pre-expansion market values.[2] The acquired homes were subsequently converted into airport-owned rental units. In other words, the number of people affected by aircraft noise did not decrease, but the previous homeowners did not incur a financial loss.

Technological Improvements for Noise Reduction

With respect to active noise abatement, one might distinguish between aircraft and engine design on the one hand and operational procedures on the other hand. The options for improving aircraft and engine design are multifaceted. A major source of engine noise is the turbulent mixing of the hot jet exhaust with the surrounding ambient air. One example for technical options that reduce engine noise are engine nacelle chevrons, a visually distinctive element on several Boeing aircraft such as 737 MAX, 787 and 747–8. They are integrated into the trailing edge of the nacelle and have been shown to significantly reduce jet-induced noise.[3]

In some cases, also later aerodynamic modifications can mitigate aircraft noise. A well-documented example is the approach noise reduction retrofit for earlier Airbus A320 models. During final approach, this aircraft emits a characteristic high-frequency whine, which is caused by airflow over circular fuel tank vents located on the underside of the wings. To address this, vortex generators (also known as airflow deflectors) – small tabs mounted upstream of these vents – are installed to generate small vortices that disrupt the airflow pattern. This modification has proven highly effective: testing campaigns have shown noise reductions of 2 to 6 dB at distances of 10 to 17 km from the runway.[4] In the meantime, many A320 operators have retrofitted their fleets with vortex generators.

Overall Development

Since the introduction of the first commercial jet aircraft in the 1950s, technological progress has decreased sound emissions per aircraft movement significantly. However, as already mentioned in Chapter 5, sound emissions rise with increasing take-off mass, therefore any comparison between sound emissions of different aircraft has to be adjusted for aircraft mass.

There are several ways to illustrate technological improvements in the area of active noise abatement. One option is a comparison of the contour area in which a certain noise level is exceeded during departure and/or arrival. For example, Embraer (2021) states that the "noise footprint area" of a modern E190 is up to 68% smaller than the one of its first-generation predecessors. Another option is to compare certified noise levels expressed as Effective Perceived Noise Level (EPNL, see Chapter 4) for aircraft of similar size, measured by maximum take-off mass (MTOM), over time. Figure 6.1 demonstrates that certified noise levels have decreased across all weight categories as aircraft technology has evolved.

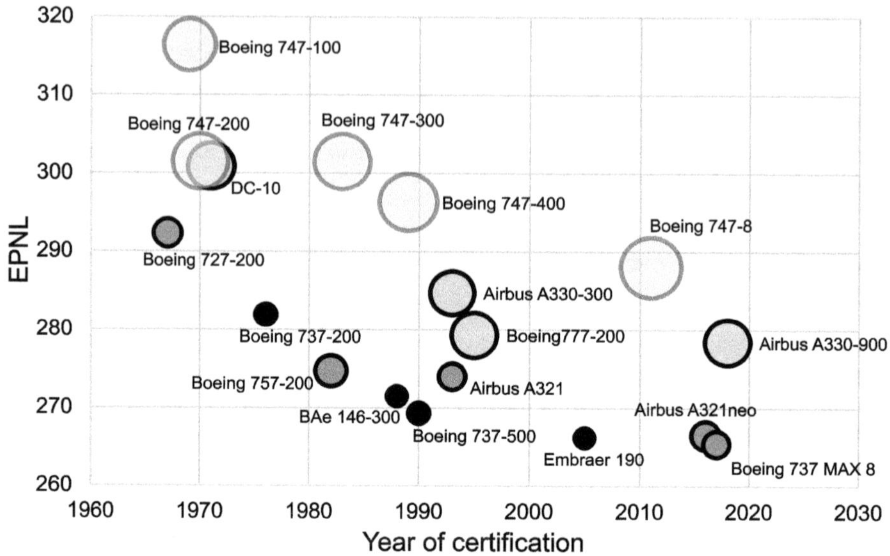

Figure 6.1 Noise levels of selected aircraft according to four weight classes and year of type certification. Circle sizes indicate MTOM, circle shadings indicate aircraft of roughly the same weight class. Figure created using data from EASA (2025b).

Operations

As already mentioned in Chapter 5, several airports mandate specific procedures for reducing sound levels on the ground or to minimise the number of residents that are severely affected. The most important options are

- Noise Abatement Departure Procedures (NADP),
- Continuous Climb Operations (CCO),
- Continuous Descent Operations (CDO),
- Adjusted Approach Angles and Displaced Thresholds,
- Low-Drag / Low-Power Configurations,

which will be described in detail in the following.

Noise abatement departure procedures are highly airport specific and depend on factors such as runway configuration, surrounding terrain, and population density near the airport. The International Civil Aviation Organization (ICAO) distinguishes between two standard procedures aimed at reducing community noise exposure during take-off: NADP 1 and NADP 2.[5]

With both procedures, thrust is reduced after the initial climb. NADP 1 is designed to minimise noise in the immediate vicinity of the airport by maintaining a steep climb at a lower speed until a defined altitude is reached. Only then is the climb rate reduced, and the aircraft allowed to accelerate, enabling flap retraction. Thus, the aircraft reaches a higher altitude while still being close to the airport. In contrast, NADP 2 reduces thrust and the climb angle earlier, which leads to acceleration and flap retraction closer to the airport. Although this means that the aircraft remains at a lower altitude when still near the airport, it is already in a "clean configuration" (flaps retracted and therefore generating less noise) and flies at a higher speed over areas at greater distances (compared to NADP 1), resulting in reduced noise exposure there. See Figure 6.2 for a visual comparison of both departure procedures.

While the primary objective of NADPs is noise reduction, their influence on vertical flight profiles – such as thrust settings, climb gradients, and flap retraction schedules – can also impact fuel consumption and emissions, leading to potential trade-offs. In addition, many modern procedures include lateral track deviations to avoid overflying densely populated areas, which increases flight distance and fuel burn but reduces population noise exposure.

Continuous Climb Operations (CCO) are another procedure for reducing the environmental impact of aircraft departures. ICAO (2013) defines CCO as "an operation, enabled by airspace and procedure design and ATC [air traffic control], in which a departing aircraft climbs without interruption, to the greatest possible extent, by employing optimum engine thrust and climb speeds until reaching cruise level". Thus, it avoids the conventional stepped climb profile with level segments and not only reduces fuel burn and CO_2 emissions but also minimises noise exposure for communities near airports.[6] Nevertheless, implementation remains challenging, particularly in congested terminal airspace or where conflicting flight paths limit climb continuity. Successful deployment depends on close collaboration between ATC, airlines and airports.

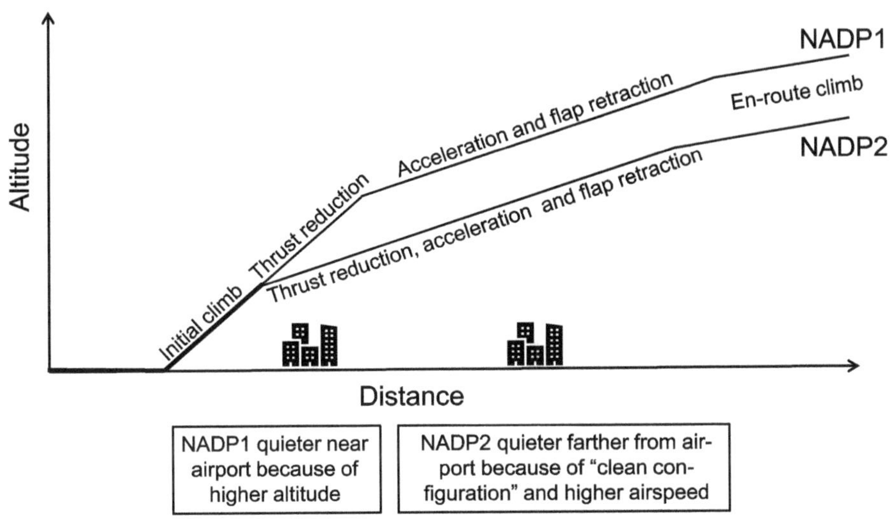

Figure 6.2 Illustration of noise abatement departure procedures (NADP 1 and NADP 2).

According to ICAO (2019), Continuous Descent Operations (CDO) enable aircraft to descend from cruise altitude to the final approach in a smooth, uninterrupted profile using minimum engine thrust and ideally low drag. This technique minimises fuel consumption and reduces both noise and emissions compared to conventional stepped descents requiring intermediate thrust. From an environmental perspective, the benefits of CDO are considerable.[7]

Beyond emissions, noise exposure for communities around airports is also significantly reduced, as aircraft remain at higher altitudes for longer and operate at idle thrust. Operational workload can be lowered as well, particularly when CDO is embedded within Standard Terminal Arrival Routes (STAR), allowing for more predictable and efficient sequencing.

Nevertheless, similar to CCO, full implementation of CDO may not be possible for a number of reasons. In high-density terminal areas, continuous descent may not always be feasible due to constraints on airspace capacity, the need to ensure safe separation between aircraft or meteorological factors requiring level flight segments.[8] Because of its benefits, many major European airports have implemented operational CDO initiatives, as documented in Eurocontrol's (2025c) CCO/CDO performance dashboard.

Another strategy to mitigate noise is the use of steeper glide slopes during precision approaches. While the standard Instrument Landing System (ILS) glide path is typically set at 3°, some airports have implemented steeper angles. This adjustment raises the aircraft's altitude, increasing vertical separation from noise-sensitive areas.[9]

A complementary measure is the use of displaced thresholds or secondary aiming points which effectively move the touchdown point farther down the runway. This increases the lateral distance between descending aircraft and residential zones located near the runway ends, particularly benefiting communities affected during night-time operations.[10]

Figure 6.3 shows an illustrative comparison between a traditional stepped approach (standard 3° glideslope) and an idealised continuous approach that is steeper and targets a displaced threshold. As illustrated, the aircraft remains at a higher altitude over the approach path, thereby decreasing noise levels in (noise-sensitive) areas close to the airport.

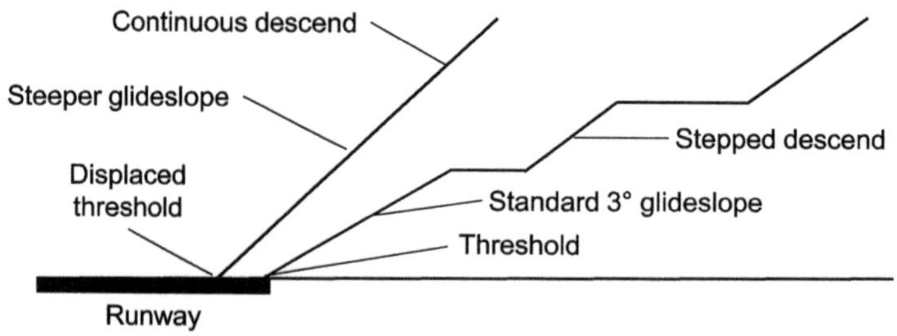

Figure 6.3 Illustration of traditional versus more efficient approach profile.

One benefit of CDO lies in maintaining a so-called low-drag, low-power configuration during the approach phase, which minimises both engine and airframe noise. Deploying high-lift devices such as flaps and landing gear increases aerodynamic drag and typically requires higher engine thrust, resulting in more noise and also a higher fuel consumption. Delaying the extension of these components can therefore significantly reduce the environmental impact. The UK Civil Aviation Authority (CAA) reports that selecting a reduced landing flap setting can cut fuel burn by up to 15 kg for an A320-class aircraft on approach, while reducing noise by up to 1.5 dB, especially during the final few nautical miles before landing.[11] Similarly, avoiding early deployment of landing gear can prevent noise increases of up to 5 dB. However, these techniques require higher approach speeds, longer landing distances and more precise control during descent, particularly under adverse weather conditions or at constrained airports.

Noise Caused by Ground Operations

Finally, airports might implement various physical structures to reduce the propagation of aircraft noise on ground. These measures aim to deflect, absorb or contain noise, particularly during taxiing, engine run-ups and the take-off roll. Typical examples are:[12]

- Ground-based noise barriers, such as walls, bunds or berms, often constructed from sound-absorbing materials or shaped with ridges and grooves to scatter sound waves,
- Engine test enclosures and noise-shielded hangars, which contain engine noise during maintenance operations or idle runs.

The effectiveness of these installations depends on factors such as their size, shape, materials and positioning. However, they also come with limitations, including high construction and maintenance costs, limited effectiveness against airborne noise and potential visual impacts. In some cases, physical barriers may also conflict with land-use constraints or aviation safety regulations.

In addition, individual noise protection is essential for airport personnel working in high-exposure zones, where the use of personal protective equipment (PPE) – such as earplugs or noise protection headsets – is typically required by safety standards.

Emissions – Local Air Quality

As shown in Chapter 4, local air quality is negatively affected by aircraft operations during the Landing-Take Off (LTO) cycle as well as by stationary sources and vehicles on the ground. The most obvious option for reducing harmful aircraft emissions on the ground is engine technology. As already mentioned in Chapter 5, emission standards have existed since the 1970s. However, similar to aircraft noise, increasing aircraft mass as well as engine thrust in general lead to higher emissions in absolute terms. Moreover, modern aircraft-engine combinations might even emit more NO_x than older aircraft with similar capacity due to the trade-off between fuel efficiency and NO_x emissions.

Operational Options

Furthermore, several operational options exist for reducing aircraft emissions. Some of these concepts have already been explained above, as they also reduce noise pollution (e.g., CDOs). Additional measures that especially target gaseous emissions during taxiing will be explained in the following.

Single-Engine Taxiing

Single-Engine Taxiing (SET) is an increasingly adopted operational measure aimed at reducing ground emissions and fuel consumption. By shutting down one engine during taxi operations modern twin-engine aircraft can significantly lower their fuel use without compromising safety.[13] SET is most suitable during taxi-in operations, as taxi-out phases often require full engine warm-up and can pose greater challenges regarding thrust asymmetry and cockpit workload. Zurich Airport (2017) estimates a net fuel burn reduction of 30–40%, both for two and four engine aircraft. While the environmental benefits are clear, the application of SET is limited by various operational factors such as airport layout, engine warm-up requirements and weather conditions, which must all be considered when defining airline-specific taxi procedures

Potential Future Concepts: Electric Taxiing and Operational Towing

Jet engines are designed to deliver thrust at high speeds and altitudes, therefore they operate very inefficiently on the ground while taxiing. A more efficient alternative would be to power the wheels of the aircraft directly (on-board solutions). Trials are already underway with electric taxiing systems that use an electric motor installed either in the nose gear or in one of the main landing gears. Key considerations include the source of electrical power, whether supplied by the APU, dedicated batteries or potentially even a fuel cell. The main drawback is the additional system weight, which makes such solutions most relevant for short- and medium-haul operations where taxiing accounts for a relatively large share of the total block time.

An alternative concept is so-called operational towing, in which tugs move the aircraft between the parking position and the runway (on-ground solutions). This approach requires no modifications to the aircraft itself, although the pilot must remain in control, potentially via remote operation. Tugs can be powered by a combustion or electric engine, making taxiing far more efficient than using the main aircraft engines. Operational challenges include determining how and where tugs manoeuvre before and after performing their task.

While both approaches reduce main engine running time, minimum warm-up and cool-down periods of the aircraft's engines must be considered, partially limiting potential efficiency gains. Additional operational factors such as pilot workload and responsibilities, air traffic control coordination and troubleshooting procedures also need to be considered. More information about the potential future taxi operations can be found in Eurocontrol's (2024) report on Sustainable Taxi Operations.

Stand Allocation and Taxi Routing

Optimising gate and taxi procedures can significantly cut fuel consumption on the ground. Eurocontrol (2024) highlights that streamlined taxi procedures drastically reduce fuel use and emissions. However, it is not useful to minimise taxi times through the allocation of the nearest runway if this means that after take-off the flight requires significant additional track miles to reach the destination.

Intersection Take-offs and Optimised Take-off Techniques

Intersection take-offs allow aircraft to begin their take-off roll from a designated point along the runway rather than from the full-length threshold. Depending on the taxi route, this can reduce the distance for taxiing and runway occupancy time, thereby contributing to a higher throughput at complex or congested airports.

However, the shorter available runway length may require higher thrust settings or impose take-off weight limitations, potentially offsetting the environmental benefits. Koudis et al. (2017), for example, estimate that reduced-thrust take-offs can lower fuel consumption and emissions by between 1% and 23% during the take-off roll, depending on the specific aircraft-engine combination.

Rolling take-offs – where the aircraft enters the runway and immediately begins its take-off roll without coming to a complete stop – can further enhance operational efficiency. This procedure avoids the high fuel flow associated with static engine run-up, slightly reduces noise exposure in the vicinity of the airport and shortens runway occupancy time.

Auxiliary Power Unit (APU) Substitution

During ground operations, aircraft frequently rely on the Auxiliary Power Unit (APU) to provide electricity to essential systems such as cabin air conditioning and lighting. However, APUs are relatively inefficient and emit substantial amounts of CO_2 and pollutants. According to ICAO (2015), depending on aircraft type and operating conditions, APU-related CO_2 emissions range from approximately 300 to 800 kg per hour. To reduce emissions, an increasing number of airports offer fixed Ground Power Units (GPU) and pre-conditioned air systems at parking positions, enabling aircraft to shut down their APUs shortly after arrival and postpone reactivation before departure. According to Zurich Airport (2018), the substitution of APU by diesel-powered GPU can reduce CO_2 emissions by more than 90% and more than 99% when switching to a fixed energy system.

Airport Collaborative Decision Making and Total Airport Management

Airport Collaborative Decision Making (A-CDM) is a procedural framework aimed at enhancing information exchange among key airport stakeholders, including airlines, air traffic control, ground handlers and airport operators. By sharing real-time data on turnaround processes and coordinating key milestones, A-CDM enables more precise planning of engine start-up, pushback and taxi clearance. This reduces unnecessary engine run time and minimises stop-and-go taxiing, leading to lower fuel consumption and emissions.

According to Eurocontrol (2016), A-CDM can reduce average taxi-out times by 0.25 to 3 minutes per departure, resulting in ground-based fuel savings of up to 7.7%. IATA (2018) therefore promotes A-CDM as a best practice, not only for improving punctuality, but also for enhancing environmental sustainability.

An extension of A-CDM is Total Airport Management (TAM), which extends the collaborative approach from day-of-operations coordination to a more comprehensive, system-wide management of airport operations.[14] While A-CDM primarily focuses on real-time process coordination, TAM integrates airside and landside activities across longer planning horizons and broader operational scopes. By placing greater emphasis on the pre-tactical planning phase, TAM improves predictability and coordinated decision making, which reduces delays and inefficient recovery measures. By addressing airport operations as an interconnected system rather than isolated processes, TAM can therefore support more efficient resource use and contribute to lower fuel consumption and emissions at airport level.

Non-Aircraft Related Emissions

Ground Vehicle Operations

At airports, a diverse fleet of airside vehicles (e.g., pushback tractors and buses) is operated by different stakeholders (especially airport operators, Ground Handling Service Providers, airlines and governmental bodies). Traditionally powered by diesel or gasoline, these ground service equipment (GSE) units contribute to local air pollution and greenhouse gas emissions.

To reduce their environmental impact, many operators are transitioning their GSE fleets to electric or other alternative propulsion technologies. Currently, the focus is mostly on CO_2 emissions (see below), but the replacement of combustion engines also reduces the emission of pollutants like NO_x. For example, Stuttgart Airport (Germany) began electrifying its GSE in 2018, including baggage and cargo tow tractors, and achieved an 80% reduction in GHG emissions in airside handling operations compared to 2009.[15] Similarly, Brussels Airport replaced its bus fleet with 30 electric buses, resulting in estimated annual reduction of 600 tonnes of CO_2.[16] Electric GSE can reduce emissions during operations by up to 100% when powered with electricity from renewable sources. They also offer additional benefits such as reduced noise. However, challenges include ensuring sufficient charging infrastructure and managing peak energy demands during operations. Hydrogen-powered vehicles are still rare at airports due to tank size constraints and infrastructure requirements. Moreover, both electric and hydrogen-based GSE conversions involve substantial capital expenditures and operational adjustments, posing considerable financial and organisational challenges for airport operators.

Beyond cleaner propulsion technologies, airports can further reduce emissions and operational inefficiencies by minimising vehicle traffic on the apron. For example, the installation of underground hydrant fuelling systems instead of using fuel trucks can lower the number of vehicle movements and associated emissions. However, constructing (and operating) such systems also causes emissions which might be analysed within a life cycle approach. In general, the deployment of advanced fleet management and routing systems enables more efficient coordination of ground service vehicles, helping to reduce idle time, optimise driving routes and limit redundant operations.[17]

Landside Access

Finally, aviation stakeholders might try to influence ground transport flows and the resulting emissions. Although airports do not have direct control over transport-related decisions of passengers and freight forwarders, they can try to influence transportation mode choices through infrastructure development and strategic cooperation with other stakeholders. Another area of influence is the promotion of low-emission ground access modes. By providing charging infrastructure for electric vehicles, airports can support travellers and especially employees to switch to these types of vehicles.

In parallel, many airports are intensifying their integration into public transport networks, particularly rail. Direct connections to regional train services or even high-speed rail (HSR) systems might offer passengers an attractive and often lower-emission alternative to private cars or short feeder flights, especially when complemented by a cooperation between airlines and rail service providers.[18] Beyond infrastructure provision, airports increasingly participate in regional and multimodal transport planning to reduce car dependency and coordinate surface access strategies. ICAO

(2022) notes that collaboration with local and regional authorities is becoming a standard element of airport sustainability planning.

Emissions – CO_2 and Non-CO_2 Effects

As outlined in Chapter 4, flight operations are by far the most important source of CO_2 emissions in aviation. In general, a distinction must be made between total emissions and specific emissions, i.e., total emissions divided by output metrics such as tonne-kilometres or passenger-kilometres. Between 1992 and 2023, total CO_2 emissions from aviation grew from 514 Mt[19] to 941 Mt,[20] i.e., an increase of almost 85%. However, in the same period, revenue passenger kilometres (RPK) grew worldwide from 1.929 bn to 8.232 bn,[21] i.e., they more than quadrupled. As a consequence, the specific emissions per passenger kilometre (pkm) decreased significantly.[22]

Specific emissions are affected by several factors. First of all, using more fuel-efficient aircraft will decrease absolute as well as specific emissions. The different technological options will be discussed in the following subchapters. Likewise, operational improvements, e.g., more fuel-efficient trajectories, will reduce absolute as well as specific emissions. However, there are several additional developments that affect specific fuel consumption even with given aircraft technologies and operations. First, as most of the fuel burn on a given flight does not depend on the number of passengers aboard, an increasing number of passengers on a given flight reduces specific fuel consumption. This can be achieved by a higher load-factor as well as a denser seating. As a result of market liberalisation as well as advances in revenue management, the average load factor increased over the last decades. Moreover, the growing market share of low-cost carriers with their rather high seat density contributed to a decrease in specific CO_2 emissions. Second, regional aircraft are characterised by rather high specific CO_2 emissions due to their low capacity. There is an ongoing trend to replace regional with larger narrow-body aircraft, also contributing to lower specific CO_2 emissions. Finally, the specific CO_2 emissions also depend on the route length. Specific fuel consumption is especially high on very short flights due to the disproportionately large share of energy-intensive take-off and climb phases. Therefore, a decreasing share in such short flights also contributes to a reduction in specific CO_2 emissions.

The Rebound Effect

Under otherwise identical conditions, improvements in specific fuel efficiency lead to a reduction in total CO_2 emissions. However, they also lower flight costs. This can lead to an expansion in flight capacity. As a result, total emissions may ultimately decline less than expected or even increase as a result of improvements in fuel efficiency.[23] This phenomenon is often referred to as the rebound effect and can also be observed in other (transport) markets.

Options for Reaching Net Zero CO_2 Emissions by 2050 – An Overview

There is an international consensus, both from policymakers (see ICAO resolution A41–21, defining the "long-term global aspirational goal – LTAG") as well as from the airline industry, that by 2050 net carbon emissions should be zero.[24] Given the growth forecasts for passenger and cargo traffic up to 2050, emissions would rise significantly in a business-as-usual scenario. Of course, this growth trend can in principle be influenced by policymakers (see Chapter 5), in particular through taxes on aviation (e.g., passenger taxes) and/or taxes on emissions. In addition, there are basically the following options for achieving net zero:

- First, in addition to modernising fleets, expanding infrastructure and improving operations will reduce specific emissions, e.g., by enabling more fuel-efficient routings or reducing taxi-times. However, the overall potential is limited, and the rebound effect mentioned above to some degree outweighs reductions of specific emissions.
- Second, the carbon intensity of energy consumption in aviation can be lowered. In the short to medium term, fossil fuel can be partially replaced by Sustainable Aviation Fuel (SAF). In a nutshell, SAF is either sourced from biological feedstocks, or it can be produced synthetically with electricity used as an energy source (often referred to as eSAF). In both cases, already existing CO_2 is turned into fuel and then emitted into the atmosphere. As there are carbon emissions linked to the SAF production process (e.g., in agriculture or for transporting feedstocks and fuels), the net-effect of using SAF is lower than 100%, currently the maximum potential is often stated to be 80–90%.[25] In the medium to long term, traditional combustion engines might be replaced by electric or even hydrogen aircraft.
- Third, as CO_2 emissions cannot be fully eliminated, emissions from the aviation industry can be offset elsewhere by either reducing emissions in other sectors of the economy or by removing CO_2 from the atmosphere, especially through additional natural sinks (e.g., trees and bogs) and possibly also by carbon sequestration and storage (CSS). This third option basically is the "net" element in aviation's net zero strategy.

There are several forecasts and roadmaps for reaching net zero carbon emissions by 2050, differing with respect to the amount of emissions in a business-as-usual case (baseline) as well as the shares of the three different options for reaching net zero emissions described above. Moreover, different terminologies and classifications can be used. In particular, the IATA four pillar concept described in Chapter 8 splits up the first option mentioned above into "operations" and "infrastructure", summarises aircraft efficiency improvements and advancements in SAF production under the headline "technology" and uses the term "Market-Based Measures" (MBM) for the offsetting options.

The baseline emissions shown in Figure 6.4 are calculated in a business-as-usual scenario and are approximately twice as high as in 2024 (see Chapter 4). Regarding mitigation options, Figure 6.4 shows that achieving net zero emissions depends primarily on the use of SAF, which accounts for around two-thirds of the total reduction. However, almost 20% of aviation's emissions in 2050 will have to be offset in order to reach net zero. Furthermore, there is a large degree of uncertainty with respect to all achievements, indicated by the vertical lines indicating maximum (lower point) and minimum (upper point) reductions based on the different scenarios.

The following sections describe the technological and operational measures that can be implemented to reduce the CO_2 emissions stemming from aircraft operations (including new aircraft and propulsion concepts). The specific aspects of SAFs as well as the shortcomings of offsetting are discussed in detail in Chapter 8. For reducing non-CO_2 effects there are limited options described at the end of this section.

Technological Measures

While the development of aircraft technologies is primarily driven by the aerospace industry, a general understanding of key innovations is important for assessing the environmental performance and long-term efficiency of aircraft operations.[26] In the following, the most relevant technological approaches and innovations are presented that contribute to reduced

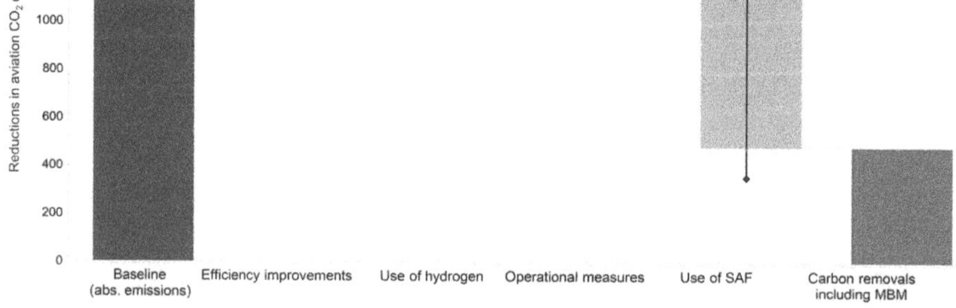

Figure 6.4 Potential of different options for reaching net zero carbon emissions in 2050 according to the IATA roadmap. Adapted from IATA (2024).

emissions and fuel burn. Such innovations are not limited to entirely new aircraft types, they are frequently introduced as production line enhancements or retrofit options for in-service fleets.

Fundamentals and Focus Areas of Technological Innovation

In straight and level flight, the four fundamental forces – lift, weight, thrust, and drag – are balanced: lift counteracts weight, and thrust equalises drag (see Figure 6.5). These relationships define the energy required to sustain flight and thereby determine an aircraft's fuel consumption.

Fuel burn – and the resulting emissions – is driven by the need to generate thrust. Consequently, three main areas of technological innovation can be identified:

- Weight reduction: A lighter aircraft has less inertia and requires less lift,[27] reducing the required thrust levels and – thus – overall fuel demand. Weight reduction applies to the airframe, its interior and the aircraft's engines. Advances in lightweight materials and structural design are central to this effort.
- Aerodynamic improvements: Minimising drag reduces the thrust needed to maintain flight. Design innovations – such as optimised wing shapes, smoother surfaces, and drag-reducing features like wingtip devices – significantly improve aerodynamic efficiency.
- Engine efficiency: Once drag and weight are minimised, reducing the fuel required to generate the necessary thrust becomes critical, and improving the thrust generated per unit of fuel is the most direct way to reduce emissions. Advanced engine designs aim to increase efficiency, delivering the same thrust with less fuel or more thrust with the same fuel input.

Together, these three areas – aircraft weight, aerodynamics and engines – form the basis for technological measures aimed at reducing the environmental footprint of aircraft operations and will be discussed in more detail in the following paragraphs.

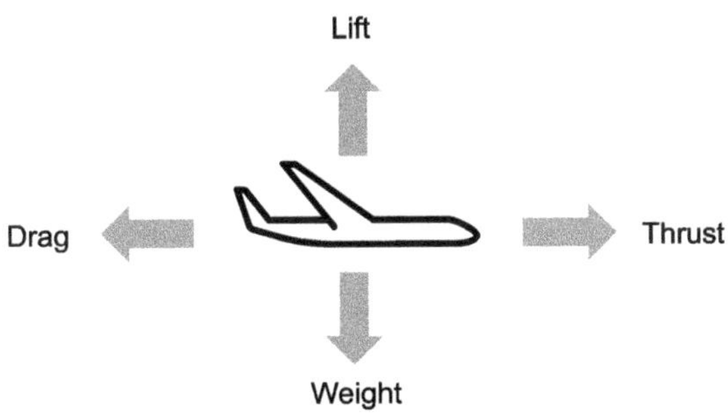

Figure 6.5. Four forces of flight: lift, weight, thrust, drag.

Lightweight Materials in Aircraft Structures

Over recent decades, aircraft manufacturers have increasingly replaced traditional metal-based structural components (primarily aluminium) with advanced lightweight materials (see Figure 6.6). The leading material among them are so-called carbon-fibre-reinforced polymers (CFRP), a composite material that offers a superior strength-to-weight ratio and can be moulded into complex aerodynamic shapes. According to Rafi (2024), composite materials allow a structural weight reduction of 15–30% compared to conventional aluminium alloys, contributing to fuel efficiency gains of up to 20–25%.[28]

These materials now dominate the structures of modern long-haul aircraft. For example, CFRP accounts for about 50% of the structural weight in the Boeing 787 and 53% in the Airbus A350.[29] Nevertheless, some critical components – such as engine pylons, landing gear elements or leading-edge structures – continue to rely on metals like titanium or high-strength aluminium alloys due to their load-bearing and fatigue-resistant properties.

Cabin and Equipment Optimisation

The use of lightweight materials is not limited to the aircraft structure.[30] Significant weight reductions have also been achieved in cabin installations such as lavatories, galleys, seat structures, and in-flight entertainment (IFE) systems. For instance, carbon-fibre panels can replace fibreglass in ceiling and sidewall assemblies, while advanced plastics and lightweight composites reduce seat and panel mass.

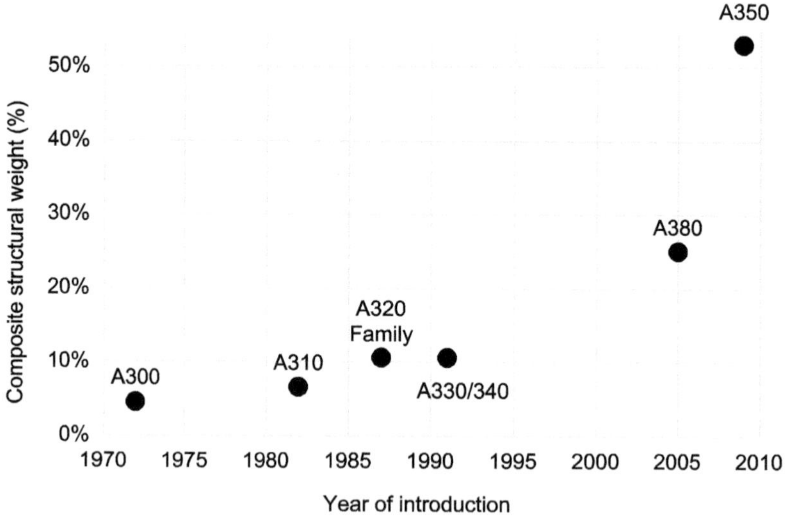

Figure 6.6 Composite material usage for Airbus aircraft. Adapted from Airbus (2015) with values approximated from original graphical representation.

Every Kilogram Counts

Operational decisions also contribute to lower weight, such as reducing the number of onboard paper-based magazines or optimising catering loads based on actual consumption data. A widely adopted innovation is the electronic flight bag (EFB), which replaces paper charts and manuals per pilot with a lightweight tablet device. While this might not sound a lot, even small weight savings can have measurable effects. According to IATA (2025b), every extra ton of weight burns/requires approximately 30 kg of fuel per hour.[31] Lufthansa Cargo's (2014) "Clear-out" programme, for instance, removed 35 kg per McDonnell Douglas MD-11 freighter by eliminating unnecessary items and replacing equipment with lighter alternatives, saving around 250 tonnes of CO_2 emissions annually.

Tankering as an Environmental Issue

Within the minima determined by safety regulations, airlines also try to minimise the amount of fuel that is carried. However, from an economic perspective there might also be trade-offs. If fuel prices differ significantly between airports, this might provide an incentive to purchase more fuel at the airport with low fuel prices (see Chapter 5 for a discussion of the potential side effects of the EU's SAF mandate).

Aerodynamic Improvements

Reducing aerodynamic drag is another strategy for enhancing aircraft fuel efficiency. Lower drag reduces the engine thrust required to maintain flight and therefore also directly decreases fuel consumption and associated emissions.

One major component of total drag is parasite drag (see Figure 6.7), which consists of three elements:

- Form drag, caused by the aircraft's shape and frontal area disrupting the airflow,
- Skin friction drag, resulting from air resistance along the aircraft's surface,
- Interference drag, which occurs at junctions between components (e.g. wing-fuselage or tailplane-empennage), where airflow is disrupted.

Parasite drag can be minimised through streamlined aircraft design. This includes smoother nose and tail contours, aerodynamically optimised transitions between fuselage and wings and the integration of components such as landing lights into flush-mounted fairings.[32] Twenty-first century aircraft such as the Airbus A350 and Boeing 787 with their more streamlined nose sections reflect these principles.[33] Composite materials also contribute to aerodynamic improvements, as

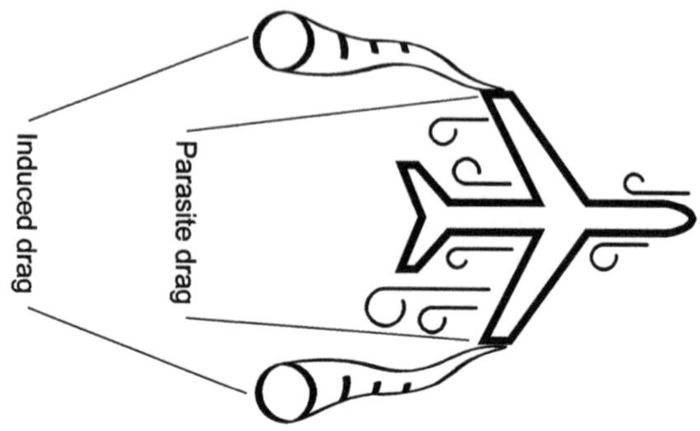

Figure 6.7. Illustration of an aircraft in flight generating parasite drag (resulting from the aircraft's surfaces) and induced drag (created at the wingtips as a consequence of lift production).

they allow large, smooth surface sections with fewer rivets or joints. Routine exterior cleaning helps maintain these aerodynamic benefits by removing surface contaminants like dirt and insects.[34]

Additional drag reduction can be achieved through microstructured surface coatings. For example, Lufthansa Technik (2025) and BASF developed a transparent riblet film inspired by shark skin. Applied to sections such as the fuselage underside, it reduces drag by directing airflow more efficiently. Flight tests with Boeing 777 aircraft have demonstrated fuel savings of up to 1%, depending on mission profile and application area.[35]

Wing Aerodynamics

While wings are essential for generating lift, they also contribute significantly to drag, particularly induced drag. This form of drag arises from the vortices that form at the wingtips due to pressure differences above and below the wing (see Figure 6.7). These vortices produce downwash, reducing aerodynamic efficiency, especially during low-speed and high-lift flight phases such as take-off and landing.[36]

Two primary design strategies exist to reduce induced drag:

- Wingtip devices: These include wingtip fences, blended winglets and raked wingtips, all of which reduce vortex strength by modifying the airflow at the wingtip. The Airbus A320 series exemplifies this development: the earliest production models had no wingtip devices, followed by the introduction of wingtip fences and now feature blended winglets (marketed by Airbus as "sharklets"). Their retrofit compatibility has made them popular across many fleets. Although they increase both weight and wingspan, such devices offer a net aerodynamic benefit especially during cruise. Depending on the design, IATA (2009) reports drag reductions of up to 5.5%.

- High aspect ratio wings: Long, narrow wings generate less induced drag than shorter, wider ones due to more favourable lift distribution and reduced vortex strength.[37] Modern aircraft such as the Boeing 777X demonstrate this principle with ultra-long folding wingtips.[38] However, high aspect ratio designs require structural reinforcement to withstand aerodynamic loads. Advanced materials like composites play a crucial role by providing the necessary strength without adding excessive weight.

Engine Efficiency

Engines are the single most significant factor influencing aircraft fuel consumption and emissions. According to IPCC (1999) turbofan engines have driven around 40% of the overall gains in jetliner fuel efficiency since the 1960s, while roughly 30% stem from airframe improvements and the rest from operations and other factors.[39]

Two main areas of technological development are relevant here: engine weight reduction and improvements in thermodynamic and propulsive efficiency.

Turbofan Engine Working Principle

A modern turbofan engine produces thrust from two distinct airflows: the core flow or primary airstream, which passes through the engine's compressors, combustion chamber and turbines; and the bypass flow or secondary airstream, which is accelerated by the large front-mounted fan and flows around the engine core. In high-bypass turbofan engines, the majority of thrust is generated by this bypass air, rather than the core exhaust.

The bypass ratio (BPR) refers to the ratio of the mass flow of bypass air to that of core air. Higher BPRs generally lead to better fuel efficiency and lower noise.

Inside the core, the compressor section consists of multiple stages, each comprising a rotating set of blades followed by stationary vanes. These stages progressively increase the pressure of the incoming air before combustion. The turbine section – located downstream of the combustion chamber – has a similar multi-stage structure. It extracts energy from the high-temperature exhaust gases to drive both the compressor and the fan via concentric shafts. Figure 6.8 provides a schematic illustration of the turbofan engine layout.

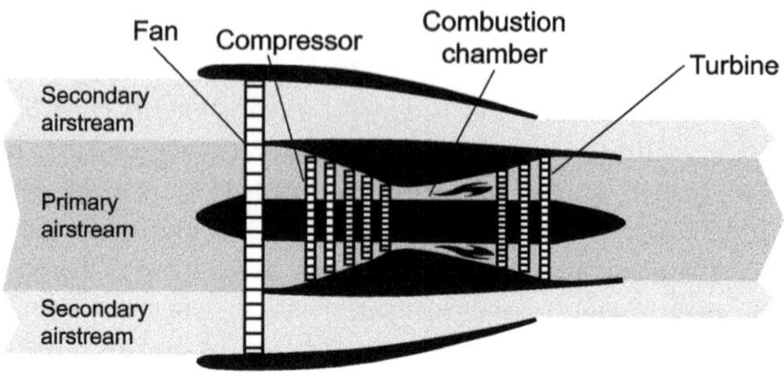

Figure 6.8. Schematic representation of a turbofan engine.

Engine Weight Reduction

Modern aircraft engines are highly complex machines weighing several tonnes.[40] A considerable share of this weight stems from the fan section, which alone can account for 20–30% of the total engine mass.[41] Reducing this mass leads to multiple benefits: not only does it decrease the aircraft's structural load, but it also reduces the required support structures in the pylon and wing mounting structures.

To reduce weight, manufacturers are increasingly turning to advanced lightweight materials such as CFRP and titanium-aluminium alloys, especially in engine fan blades and casings. As outlined in the section on airframe weight reduction, CFRP offers a superior strength-to-weight ratio and enables complex aerodynamic forms, making it well suited not only for fuselage components but also for key engine structures under mechanical and aerodynamic loads.[42]

Engine architecture has also evolved: by optimising the aerodynamic design of fan and compressor stages, manufacturers can reduce both the number of blades and the number of stages required to achieve the desired compression. Fewer, more efficient blades decrease engine weight and aerodynamic drag, while a lower stage count simplifies the engine layout and reduces weight and internal friction losses.[43]

Thrust-Specific Fuel Consumption, Thermal Efficiency and Bypass Ratio

Fuel efficiency in jet engines is commonly measured as thrust-specific fuel consumption (TSFC), the amount of fuel required to produce a given unit of thrust. One way to improve TSFC is through more efficient combustion, made possible by advanced materials and manufacturing techniques. These enable engines to operate at higher temperatures with reduced internal cooling, thereby increasing thermal efficiency. As a result, combustion and heat extraction become more effective, leading to lower fuel consumption per unit of thrust.[44]

The most significant factor, however, is the bypass ratio (BPR). A higher BPR means that more air contributes to forward thrust without being mixed with fuel. For example, in an engine with a BPR of 7:1, seven units of air bypass the core for every one unit that passes through it. The most advanced turbofan engines can reach BPRs of up to 12.5:1 (available for the Airbus A320neo). For comparison, the engine powering the A320ceo has a BPR of approximately 5.5–6:1.[45]

Larger bypass ratios require larger fan diameters, which introduces design challenges, particularly in terms of weight, aerodynamic drag and ground clearance under the wing. This issue became especially apparent on the Boeing 737 MAX, where the low ground clearance of the legacy 737 design required changes to engine placement, ultimately contributing to aerodynamic changes and certification issues.[46] Another challenge associated with larger fan diameters is the increase in tip speed at a given rotational rate. Excessive tip speeds can lead to efficiency losses and higher noise levels, particularly if they approach or exceed the speed of sound. To prevent this, fan blade tip speeds are typically kept below supersonic levels. One solution to this issue is the Geared Turbofan (GTF) developed by Pratt & Whitney (PW). By incorporating a planetary gear system that decouples the rotational speeds of the fan and the low-pressure turbine, the GTF allows for a larger fan diameter while enabling each component to operate at its optimal speed. This results in improved efficiency and reduced noise.[47] The GTF is available from PW as PW1100G, featuring a bypass ratio of 12.5:1 and is one engine option for the Airbus A320neo series. The alternative engine option is CFM's LEAP engine, which follows a conventional turbofan design without a reduction gearbox. Despite its slightly lower bypass ratio of 11:1, it achieves comparable efficiency gains, thanks to extensive use of lightweight materials, advancements in the combustion chamber and the absence of the added weight associated with a gearbox.

Potential Future Concepts: Innovative Propulsion Systems and Aircraft Designs

In the previous section, various technological measures to reduce aviation's environmental footprint were presented. Many of these can be further developed through incremental improvements, such as increasing engine bypass ratios, adopting higher aspect ratio wings, introducing more advanced aerodynamic features, using lighter materials, etc.

By contrast, the following focuses on disruptive approaches: changes in aircraft and engine technology that go beyond evolutionary optimisation of existing systems. Such innovations fundamentally alter how aircraft are powered, designed and operated.

Open Rotor Engines

Open rotor engines[48] aim to combine the efficiency of turboprops with the speed and performance of turbofans. Today, both engine types are in widespread use: turbofans are optimised for fast cruise at high altitudes, while turboprops are most efficient at lower speeds and altitudes, making them common for regional aircraft.[49]

Open rotor designs seek to merge the advantages of both concepts by having unducted counter-rotating propellers with many aerodynamically optimised blades mounted at the rear of the engine nacelle. The absence of a duct or casing reduces weight and drag while enabling ultra-high bypass ratios. Although the concept is not new – extensive trials were already carried out in the 1980s – aviation's decarbonisation targets have revived interest.[50] Currently, Safran and GE Aerospace are developing an open rotor demonstrator having a BPR "more than five times greater than tomorrow's most advanced ducted engines" and aiming to be 20% more fuel efficient.[51]

However, these well-known benefits are accompanied by equally familiar challenges.[52] While CO_2 emissions can be reduced, there is little impact on NO_x emissions, which remain largely dependent on the core engine and its internal combustion process. Moreover, integrating such engines into existing aircraft designs is difficult due to their large fan diameter, which rules out conventional under-wing mounting and instead requires high-mounted wings or rear-fuselage installation. Finally, the probably most significant challenge is the increase of overall noise levels and generation of distinct tonal frequencies that are perceived as more irritating, and therefore noisier, than those of current aircraft engines.

Battery-Electric Propulsion

Fully electric aircraft promise zero emissions during flight and a mechanically simpler propulsion architecture. They use propellers or ducted fans that are powered by electric engines receiving their energy from batteries.

The fundamental limitation of this concept lies in the battery's low energy density: 1 kg of jet fuel contains about 43 MJ specific energy while 1 kg of state-of-the-art batteries contains around 2 MJ (see Figure 6.9). In other words, to match the energy content of 1 kg of jet fuel, a battery weighing 21.5 kg would be required.[53] This leads to severe constraints on range and payload, further intensified by the fact that battery mass remains constant throughout the flight unlike fuel which is consumed and thus reduces aircraft weight over time.[54] Additional operational challenges include the need for electricity from renewable sources, as well as the time required for recharging.[55]

A fundamental breakthrough in battery technology would be required to overcome these limitations, but progress so far has been gradual and insufficient to meet the energy demands of larger aircraft.[56] Thus, given the current technological constraints, full-electric propulsion is unlikely to play a significant role in decarbonising medium- or long-haul aviation in the near term.

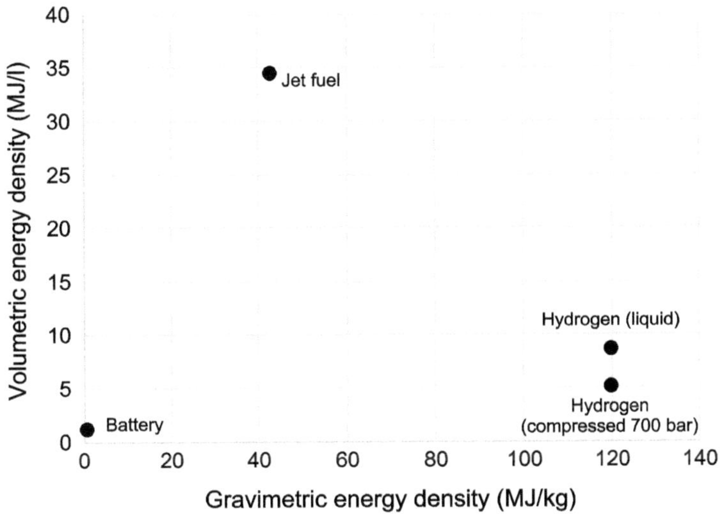

Figure 6.9. Energy density of jet fuel, hydrogen and batteries. Figure created using data from ICCT (2022) and Schmelcher and Häßy (2022).

Hydrogen-Based Propulsion

Hydrogen propulsion can be implemented in two ways: through direct combustion in modified jet engines or via fuel cells that generate electricity to power electric motors. Both approaches eliminate CO_2 emissions at the point of use and offer the potential for significant reductions in other air pollutants and contrail formation, particularly when using so-called green hydrogen.[57]

- Combustion-based hydrogen propulsion operates similarly to conventional jet engines. However, unlike hydrocarbon-based fuels, hydrogen does not contain carbon and therefore produces no CO_2 emissions. But high-temperature combustion in ambient air still results in NO_x emissions due to the presence of nitrogen. Current research focuses on quantifying these NO_x levels and also assessing contrail formation.[58] Because hydrogen contains fewer impurities (such as sulphur) than fossil fuels, it should generate fewer condensation nuclei and thus potentially fewer contrails.
- Fuel-cell hydrogen propulsion uses an electrochemical process to produce electricity from pure hydrogen and oxygen, generating only water as a by-product. Since nitrogen and other atmospheric components are excluded from the reaction, NO_x emissions are not produced. The water produced is in liquid form and could be collected or released (though, if released, it would likely form ice crystals). Because fuel cells operate most efficiently at steady power output, batteries are typically required

both as a buffer and to meet peak power demands during dynamic flight phases such as take-off. This adds weight and complexity, as the system must integrate both the fuel cells, including their cooling systems, and the battery storage.

Regardless of the conversion method, hydrogen faces two major challenges: energy density and fuel infrastructure. In its gaseous state, hydrogen contains less than 0.03% of the energy per litre of conventional jet fuel. To reach practical storage densities, it must either be compressed or liquefied, the latter requiring cryogenic storage at temperatures below -253 °C. Even then, compressed hydrogen at 700 bar provides only about 16% of the volumetric energy content of jet fuel, while liquefied hydrogen reaches roughly 25% (see Figure 6.9). Consequently, aircraft powered by hydrogen must accommodate substantially larger fuel volumes to achieve the same energy output as jet fuel.

Furthermore, compressed hydrogen requires large, structurally robust cylindrical tanks, while liquefied hydrogen must be stored in heavily insulated cryogenic tanks. In both cases, the tank structures add significant weight, thereby reducing fuel efficiency. Moreover, hydrogen storage tanks cannot be integrated into conventional wing structures and must instead be located within the fuselage, which further affects aircraft layout and reduces available payload space.[59]

And beyond the aircraft itself, a complete redesign of airport infrastructure and fuel supply chains would be necessary to enable hydrogen transport, storage, refuelling and associated safety systems.

Several initiatives exist to develop hydrogen-based propulsion systems, either as retrofits or for entirely new aircraft designs. However, similar to electric flight, these efforts currently focus on the short-haul (regional) market with small commuter aircraft. Airbus' ZEROe project, originally aiming to introduce a hydrogen-fuelled mid-size aircraft by 2035, was recently postponed. Boeing, in contrast, has so far shown little ambition in pursuing hydrogen-fuelled aircraft.

High-Aspect-Ratio Wing Designs

As described, one factor in aerodynamic efficiency is the aspect ratio: the ratio of wingspan to average chord length. Higher aspect ratios reduce induced drag, but for large commercial aircraft there are physical and engineering limits. Aircraft must be structurally capable of withstanding significant loads, including those encountered under turbulence certification standards. One proposed solution is the strut-braced or truss-braced wing, where the wings are mounted high on the fuselage and supported by external struts. This high-wing configuration also facilitates the installation of engines with large diameters like high BPR engines or even open-rotor designs.

The trade-off of this concept is additional drag and weight from the strut itself. Also – similarly to the Boeing 777X – folding wings might be required to allow compatibility with current airport infrastructure.[60]

Both large manufacturers Airbus and Boeing are exploring high-aspect-ratio solutions but with differing approaches. Boeing, in collaboration with NASA, planned to develop a full-scale demonstrator for the truss-braced wing concept. However, the project was paused in April 2025 following a shift in focus, although Boeing intends to continue evaluating alternative approaches to refine thin-wing technology.[61] Airbus (2023) is also investigating high-aspect-ratio wing designs but seeks to achieve long spans without struts, employing advanced aerodynamic surfaces and turbulence-control sensors that allow the aircraft's systems to automatically adapt the wing to prevailing conditions.

Blended Wing Body

Current aircraft layouts maintain a clear division of functions: the fuselage carries the payload (but adding weight and drag) while the wings generate lift. This arrangement is known as the tube-and-wing concept. In contrast, the Blended Wing Body (BWB) concept integrates fuselage and wings into a single, continuous lifting surface, reducing drag and improving lift efficiency. While initial estimates suggested that an optimally designed BWB could achieve up to 25% fuel savings compared to a conventional tube-and-wing configuration, more recent studies indicate an efficiency gain of approximately 15% compared to the current aircraft generation.[62]

In addition to potential aerodynamic gains, further benefits include noise reduction, as engines can be mounted on top of the aircraft, with the airframe itself shielding the ground from engine noise. BWB configurations are also particularly promising for hydrogen-powered aircraft, as they can more easily accommodate large cylindrical tanks within or above the wing structure.

However, BWB designs face several challenges. Passenger comfort can be affected by the lack of windows and by motion perception in seats located far from the centreline. Certification processes are likely to be more complex, and emergency evacuation procedures may require new solutions. Scalability is another issue: in today's tube-and-wing concept, capacity can be adjusted by lengthening or shortening the fuselage. This is far less straightforward for a BWB, and if scaled down too much, the ratio between wing area (for lift) and wing volume (for payload) becomes unfavourable. Finally, the pressure cabin is non-cylindrical, which presents additional structural challenges.

Currently, BWB concepts are tested only with small-scale prototypes and no full-scale designs from leading manufacturers have yet been realised. But Jet-Zero in cooperation with Northrop Grumman plans the first flight of a full-scale demonstrator in 2027.[63]

Operational Measures

While the previous sections focused on technological innovations – usually driven by aerospace engineering – operational measures address how existing aircraft and infrastructure can be used more efficiently. Although many of these procedures are already conducted every day and generally fall under the responsibility of operations, air traffic control (ATC) or ground handling, it is important to understand their environmental relevance from a systems perspective.[64]

The following section outlines some key operational strategies that aim to reduce fuel burn and emissions during the different phases of flight: on the ground, during take-off and climb, in cruise, and during approach and landing. Even without new technologies, these practices offer potential for environmental improvements. In particular, onboard decision-support tools integrated into modern flight management systems (FMS) help flight crews identify optimal altitudes, speeds and route adjustments in real time and, thus, enable environmentally preferable choices even under dynamic conditions.

As mentioned in the previous section on noise reduction, several operational procedures during approach and departure do not only reduce noise emissions but also fuel burn. These options will not be mentioned again in this section.

The cruise phase accounts for most of total fuel burn and CO_2 emissions of flights, making it the most impactful stage for environmental optimisation.[65]

Ignoring the influence of wind, the direct route typically represents the most fuel-efficient path for any flight.[66] However, legacy route networks – often shaped by national boundaries, military restrictions, and fragmented airspace control – continue to constrain flight planning. These structural inefficiencies can result in measurable deviations from the optimal route.[67] To address these issues, several large-scale initiatives aim to modernise airspace management and enable more direct routing, for example the Free Route Airspace initiative.

Moreover, it has to be pointed out that in some cases economic considerations might lead to increased fuel burn. In Europe, ANSPs levy en-route charges, and these charges might differ significantly between neighbouring states. Consequently, in some cases airlines might reduce costs if they fly longer routes – and therefore burn more fuel – that avoid airspaces with rather high charges.[68]

Potential Future Concept: Formation Flights

The potential for truly disruptive operational changes is limited and the only major option would be formation flying.

The Airbus (2025) "fello'fly" concept is inspired by the V-shaped flight formations of migrating birds. In nature, the wingtip vortices generated by the lead bird create an upwash that provides additional lift to the following bird, reducing its energy requirements. The same aerodynamic principle can be applied to commercial aircraft: when positioned correctly a trailing aircraft can benefit from the upwash of a leading aircraft, thereby reducing fuel burn.

Formation flying itself is not new, military aircraft have long operated in coordinated formations. Airbus has examined the technical and physical feasibility of adapting this principle to commercial aviation. Technical implementation is comparatively straightforward, requiring no major airframe modifications but mainly software updates and crew training. Initial flight tests have demonstrated significant fuel savings for the trailing aircraft.[69] While the idea might sound unconventional, theoretically it could align well with existing long-haul traffic patterns shaped by airline schedules.[70]

Nevertheless, operational, technological and business-related challenges remain. These include the need for highly precise navigation and spacing, effective wake turbulence management, seamless air traffic control coordination and the creation of viable business models for benefit-sharing between airlines. From a regulatory standpoint, current airspace management is designed to maintain minimum separation distances to avoid wake turbulence, whereas fello'fly aims to intentionally exploit it. Accommodating such operations would therefore require new regulatory provisions from ICAO and national aviation authorities.

Measures on the Airport Level

As presented in Chapter 4, the direct contribution of airports to overall aviation-related greenhouse gas emissions is relatively small, typically between 1% and 5%. In the following, an overview on mitigation measures is provided, focusing on those options that are rather specific to the airport industry. In general, one can distinguish between two complementary approaches:

- Reducing overall energy demand, through improved efficiency in airport systems, buildings, and ground operations,
- Reducing the environmental impact of energy generation, for example by switching to renewable or low-emission energy sources.

Energy Efficiency at Airports

Airports consume substantial amounts of energy across a wide range of systems, leading to CO_2 and other greenhouse gas emissions either directly at the airport or at energy providers.[71] Measures to improve energy efficiency span across terminal infrastructure, outdoor systems, and airside and landside vehicle operations, which have already been discussed in the previous section on local air quality. Although many of these approaches are common in other industries, airports have specific challenges due to their operational complexity, large scale and safety-critical infrastructure.

Lighting Systems

One of the most relevant efficiency measures is the replacement of conventional incandescent or halogen lighting with LED technology. LEDs consume up to 80–85% less energy for the same light output and offer significantly longer lifespans, reducing maintenance needs. These systems are widely implemented both inside terminals and in airside operations, including apron, taxiway and runway lighting. In addition, lighting systems can also be operated on demand, using motion or daylight sensors in infrequently-used indoor areas. On runways and taxiways, lighting can be activated as needed, for example when aircraft are approaching.

HVAC and Building Optimisation

Depending on the location and climate conditions, Heating, Ventilating and Air Conditioning (HVAC) can account for a major share of energy consumption. GIZ (2022) provide a comprehensive overview about measures to reduce the energy need for HVAC and specific measures might include:

- Thermal insulation of walls, roofs, windows and doors, which can reduce heating energy demand by up to 90% and cooling electricity demand by up to 50% compared to non-insulated structure,
- Sun protection systems such as exterior blinds or shading elements, which can cut cooling-related electricity use by approximately 33% on average,
- Occupancy-based control systems using sensors to adapt temperature and ventilation settings to real-time use, resulting in energy savings of up to 20%,
- Data analytics to forecast and optimise energy needs based on weather conditions, passenger volumes, or operational schedules.

To support these measures, advanced energy management and information systems are increasingly deployed to monitor, analyse, and optimise building energy consumption in real time.

Energy Supply at Airports

Most airports source their electricity and heating from off-site energy providers. The environmental footprint of this energy consumption depends heavily on the local or national energy mix, just as it does for households and businesses/factories across the respective region. Although airports do not control how this energy is generated, they can influence its composition through strategic procurement. This includes entering into long-term power purchase agreements with renewable energy providers or forming partnerships with suppliers that prioritise low-carbon energy sources. To name one example, Fraport (the owner/operator of Frankfurt Airport) and the energy provider supplier EnBW have concluded a long-term power purchase agreement for the supply of electricity generated by offshore wind turbines in the North Sea off the coast of Germany.[72]

On-Site Renewable Energy Generation

In addition to sourcing energy from external providers, a growing number of airports install on-site renewable energy systems to reduce emissions and increase energy autonomy. The most widely adopted technology is photovoltaics (PV), as airports are particularly well suited for such installations due to their extensive rooftop areas and unused land between runways, taxiways and buildings. At Frankfurt Airport, even vertically mounted PV panels have been installed in confined spaces near the runway.[73] Although less efficient due to limited sun exposure, these systems still contribute to decarbonisation goals. While solar panels are a relatively straightforward and low-maintenance technology, their use at airports requires specific aviation-related considerations – most notably the need to prevent glare near operational areas.[74]

Beyond solar, some airports have adopted geothermal heating systems. For example, Zurich Airport (2023) is exploring the viability of underground thermal storage using a subglacial channel located 300 metres beneath the airport's surface. Waste-to-energy solutions that convert organic or residual waste into heat are also gaining attention, with Gatwick Airport often cited as a best-practice example.[75] Wind turbines, by contrast, remain rare due to height restrictions and the risk of turbulence, making them viable only in small-scale or specially designed applications. Nonetheless, the UK's East Midlands Airport (2020) has already installed commercial-grade wind turbines on-site.

Options for Reducing Non-CO_2 Effects

As presented in Chapter 4, CO_2 emissions account for only about one-third of the total climate impact of aviation. A major contributor to the non-CO_2 effects is the formation of contrails and contrail-induced cirrus clouds (also referred to as aviation-induced cloudiness, AIC), which form when water vapour emitted by aircraft freezes in Ice-Supersaturated Regions (ISSRs).

One mitigation pathway involves the use of Sustainable Aviation Fuel (SAF), which generally produce fewer soot particles and thus fewer condensation nuclei, potentially reducing both the likelihood and radiative impact of contrails (see Chapter 8).

Moreover, operational strategies are also being explored, such as avoiding contrail-prone airspace (i.e., ISSRs). This typically requires horizontal or vertical rerouting, which may lead to longer flight paths, less efficient cruising altitudes or increased airspace congestion. These factors can in turn raise CO_2 emissions, partially or fully offsetting the intended climate benefit. In addition, forecasting ISSR locations remains a significant meteorological challenge, limiting the predictability and effectiveness of such avoidance strategies in real-time operations. Due to these uncertainties, the mitigation of contrails and AIC remains a key area of research in aviation climate science.[76]

Other Environmental Aspects of Airport Operations

In addition to emissions and energy use, airports are also associated with other environmental impacts, particularly in the areas of soil and water contamination, land consumption and waste management (see Chapter 4). These aspects are often site-specific and governed by local environmental regulations.

Water and Soil Contamination

As mentioned in Chapter 4, the two main sources of water and soil contamination at airports are rubber abrasion during aircraft landing and the use of de-icing and anti-icing fluids. Rubber particles accumulate especially in runway touchdown zones. While runways are regularly cleaned to ensure adequate friction, surrounding grass or soil areas retain rubber debris that is more difficult to remove and may leach into the environment over time. Although this specific pathway is less addressed in regulatory frameworks, the accumulation and dispersion of pollutants is discussed in ICAO's (2021) environmental guidance material "Water Management at Airports". It emphasises the need for integrated stormwater and surface pollution management strategies that address runoff, infiltration and the long-term protection of groundwater resources.

Glycol-based de-/anti-icing fluids pose a significant risk to aquatic ecosystems due to their high biochemical oxygen demand. Since there are no operationally equivalent and environmentally neutral alternatives, airports rely on mitigation measures such as minimising fluid use and capturing runoff. One key strategy is the construction of dedicated de-icing pads located close to runways. These facilities often feature sealed surfaces, drainage systems and underground storage tanks, which allow for the collection and treatment – or even recycling – of glycol-contaminated runoff.[77]

Stankevičienè et al. (2019) report that more advanced airport facilities employ technologies such as anaerobic bioreactors, filtration systems, or constructed wetlands for the treatment of contaminated runoff. Munich Airport, for instance, operates an on-site treatment plant that recovers and recycles glycol from de-icing fluids. Other airports, including Toronto Pearson, Kalmar (Sweden) and Heathrow (UK), have adopted wetland-based systems that rely on natural subsurface processes to purify runoff.

Waste Management

Airports generate substantial quantities of waste from terminal operations, catering services, aircraft cabins, retail concessions and maintenance activities. Effective waste handling is essential for environmental protection and public health, especially given the scale and diversity of waste streams involved. ICAO (2018) addresses this topic in its "Guidance on Airport Waste Management", which presents best practices for waste prevention, material separation, recycling and final disposal. It emphasises the importance of integrated waste strategies tailored to each airport and regulatory environment, while also promoting the adoption of circular economy principles where feasible. To manage these waste streams, many airports implement separation systems for recyclable materials such as paper, plastics, metals, glass and organic waste. In several jurisdictions, e.g., in the EU, these efforts are supported by regulatory measures such as the ban on certain single-use plastics and incentives for higher recycling quotas.[78]

However, international catering waste remains a major challenge. Classified in many countries as so-called Category 1 high-risk waste, it must be burnt regardless of its recyclability due to strict biosecurity regulations. As discussed in Chapter 4, this limits material recovery and increases emissions. A notable example – previously mentioned in the context of on-site energy generation – is London Gatwick Airport, operating an integrated waste-to-energy facility. It processes up to 10 tonnes of residual and Category 1 waste daily, converting it into biomass fuel that supplies renewable heat to the airport's North Terminal. Complementary technologies, such as drying units, separation systems and baling presses, have further reduced the number of waste-related vehicle movements on-site, thereby lowering CO_2 emissions and easing apron traffic.[79]

Land Use and Surface Sealing

As discussed in Chapter 4, airport infrastructure requires large areas of land and leads to extensive surface sealing, which can have a negative impact on soil penetrability, the local water cycle and surrounding ecosystems. These effects are inherent in the construction of runways, taxiways, aprons and terminal buildings and are generally unavoidable due to strict safety requirements, among other things. Although minor optimisations are possible in the planning stage, reducing the surface area – for example, by narrowing the taxiways or reducing the apron space – would lead to operational bottlenecks.

In order to mitigate the environmental impacts associated with land use, some airports are required by national or regional environmental regulations to provide ecological compensation areas. These may include the restoration of wetlands, the creation of grassland habitats or the reforestation of land elsewhere in the region to maintain biodiversity and ecosystem functions. A notable example is Frankfurt Airport, where the extensive compensation measures had to be taken for the 2011 expansion, including the reforestation of the cleared forest on the same scale.[80]

Green roofs are increasingly being used on existing buildings as a measure to partially restore environmental functionality. In addition to thermal insulation and noise reduction, green roofs can also reduce rainwater runoff and mitigate the heat accumulation effects similar to the urban heat island phenomenon, which can occur in large, sealed airport areas.[81] However, their implementation in the airport context is often constrained by structural weight limits and potential conflicts with photovoltaic (PV) systems, which are often preferred due to their ability to contribute to both environmental and economic performance goals. The choice between green roofs and solar installations must therefore be made carefully based on site-specific priorities and technical feasibility.

Key Takeaways

- Environmental mitigation in aviation is inherently multidimensional, as measures targeting noise, local air quality and climate impacts often generate synergies but can also create trade-offs, particularly between fuel efficiency and pollutant emissions such as NO_x.
- Substantial improvements in aircraft noise mitigation have been achieved over recent decades through advances in engine design, aerodynamics and operational procedures. However, increasing aircraft size and traffic volumes partly offset per-movement noise reductions.
- Operational measures, including optimised departure and approach procedures (e.g., CCO, CDO, NADP), can simultaneously reduce noise, fuel burn and emissions. But their widespread implementation is constrained by airspace capacity, safety requirements and coordination necessities among stakeholders.
- Local air quality around airports can be significantly improved through ground-operation measures, such as single-engine taxiing, APU substitution, optimised taxi routing and electrification of ground service equipment. Operational feasibility and infrastructure investments remain limiting factors.
- Airport-specific environmental impacts (e.g., water and soil pollution, land use) remain relatively unabated, and, thus, call for site-specific environmental management approaches.
- Technological improvements in aircraft efficiency, driven by weight reduction, aerodynamic optimisation and improved engine performance, have significantly reduced specific CO_2 emissions. However, absolute emissions have grown due to continued demand growth and rebound effects driven by lower operating costs and fares.

- Net zero CO_2 emissions by 2050 cannot be achieved through efficiency gains alone, making large-scale deployment of Sustainable Aviation Fuel (SAF) the dominant short- to medium-term mitigation pathway, complemented by offsets for residual emissions.
- Non-CO_2 climate effects, especially by contrails, remain a significant challenge that requires further research and targeted operational strategies.
- Disruptive propulsion and airframe concepts – such as open-rotor engines, battery-electric aircraft, hydrogen propulsion and blended wing bodies – offer long-term mitigation potential, but face substantial technological, operational and regulatory barriers that confine their relevance primarily to specific market segments in the foreseeable future.

Notes

1 For example, Freeman et al. (2018) analyse the trade-off between CO_2 and NO_x emissions with respect to global warming.
2 For this so-called CASA program see Fraport (2011).
3 IATA (2009).
4 GreenAir (2015).
5 Hoyas (2025).
6 Using 2017 data, Eurocontrol (2025a) reports per-flight savings of approximately 15 kg of fuel for CCO-departures in Europe.
7 A study by Eurocontrol (2025a) found average fuel savings of 35 kg per arrival, corresponding to roughly 110 kg of CO_2, when comparing CDO with standard stepped approaches.
8 Eurocontrol (2025a) reports that while 41% of arrivals achieved a continuous descent from flight level 75 (7,500 feet), only 24% could begin CDO from cruise level.
9 According to CANSO (2015), an intercept angle of 3.2° keeps the aircraft at higher altitudes resulting in additional noise reductions of 3–5 dB between approximately 30 km (19 NM) and 8 km (5 NM) from the runway.
10 See CAA (2014) for more details.
11 CAA (2014).
12 CAA (2014).
13 The concept also applies to multi-engine aircraft, where more than one engine can be shut down during taxiing.
14 Eurocontrol (2025d).
15 These numbers refer to emissions during operations. Including life cycle emissions, e.g., for producing batteries, probably leads to different results.
16 These examples are presented by GIZ (2022) in its report "Energy Efficiency in Airports", which offers a comprehensive and detailed overview of a wide range of measures implemented by or available to airports.
17 See for example Sunaryo and Zuniga (2025) for further details.
18 While the integration of improved transit access and especially HSR services can reduce surface emissions and road congestion, it also expands the airport's catchment area, potentially leading to additional air traffic (demand) with its environmental consequences.
19 IPCC (1999).
20 IEA (2024a).
21 ICAO (2024) and previous ICAO publications.
22 Both emission values include military aviation, but there is no precise information available on potential changes in the share caused by the military. Emissions from

civil aviation include those from all cargo flights as well as emissions caused by the transport of belly cargo. Therefore, simply dividing total emissions by passenger kilometres is not appropriate. However, as the by far largest share of emissions result from passenger transportation, it is correct to state that specific emissions decreased significantly.

23 For an early analysis see Evans and Schäfer (2013).

24 IATA (2025a) and IATA (2021).

25 While SAF produced along current technological pathways enables a CO_2 emissions reduction in the range of 80% (compared to fossil fuel), so-called advanced SAFs, especially eSAFs, in the future can achieve life cycle emissions reductions of 90% and even more. See for example ICCT (2025).

26 The "Technical Annex" of IATA's 2009 "Technology Roadmap" provides a comprehensive overview and detailed explanations about technological measures to reduce the environmental impact of aircraft operations (IATA 2009).

27 Less lift also reduces induced drag, a by-product of lift generation. It results from the vortices formed at the wingtips due to pressure differences between the upper and lower wing surfaces. This type of drag increases with the amount of lift produced.

28 However, from a life cycle perspective such materials might also have disadvantages like a higher energy consumption for production or for recycling.

29 Dubois (2020).

30 IATA (2009).

31 Sources report significantly different values. For example, Burden (2025) states that a reduction of 1 kg in weight saves approximately 0.003 kg of fuel per kilometre, while Steinegger (2017) reports a saving of only 0.02–0.03 kg per 1,000 km, an order of magnitude lower.

32 For example, on the A320neo, Airbus offers the so-called MFRL package (Multifunctional Runway Lights), a set of LED lights integrated into the wing root. This aerodynamically cleaner installation replaces the landing lights and other lights that were previously mounted on the nose gear strut and exposed to the airstream, thereby generating additional drag.

33 This also includes the reduction in the number of lateral cockpit windows from four to two (Kingsley-Jones 2005). Even small details, such as vertically stowed windshield wipers, are optimised for improved airflow. According to Poland (2020), such measures can reduce drag by around 1%, making them attractive even for retrofit applications in the industry.

34 According to Avia Solutions Group (2022), a freshly cleaned aircraft can achieve fuel savings of approximately 0.5%, a figure commonly cited in the industry.

35 For a fleet of 10 aircraft, this corresponds to annual savings of approximately 4,000 tons of fuel and nearly 13,000 tons of CO_2 emissions, equivalent to around 53 individual freight flights from Frankfurt to Shanghai. The measure is currently being implemented within the Lufthansa Group and has also been adopted by other carriers.

36 While parasite drag affects any object moving through air (e.g., also cars or bicycles), induced drag is unique to lifting surfaces like aircraft wings. It also contributes to wake turbulence, which poses a hazard to following aircraft.

37 A good example is provided by gliders, which are inherently designed for maximum aerodynamic efficiency. One of their most defining features is the use of long slim wings.

38 The wingtips are folded upwards on the ground, reducing the overall wingspan. This ensures compatibility with existing airport dimension restrictions while still allowing the aircraft to benefit from improved aerodynamics in flight.

39 For example, according to Boeing (2025), the 787 is up to 20–25% more fuel efficient than the aircraft it replaces (such as the 767 or early 777 models). The

787's engine alone is about 15% more efficient than its predecessor (Aslam 2025), meaning that a substantial share of the overall efficiency gains – approximately 15 percentage points out of the total 25 – can be attributed to the engine.

40 According to its type certification by EASA (2022), General Electric's GEnx engine – used on the Boeing 787 and 747–8 – has a dry weight between 5.6 and 6.1 tonnes.

41 IATA (2009).

42 IATA (2009).

43 For example, General Electric's GE9X engine for the Boeing 777X uses only 16 composite fan blades, compared to 22 on its predecessor, the GE90.

44 See IATA (2009) and ATAG (2024). For example, the GE9X engine incorporates heat-resistant ceramic matrix composites and 3D-printed fuel nozzles, enabling hotter combustion and improved fuel-air mixing. This enhances overall efficiency while reducing NO_x emissions.

45 Aircraft Commerce (2019).

46 The Boeing 737 MAX design placed the larger engines further forward and higher on the wing to maintain ground clearance. This change in geometry contributed to altered flight characteristics and the need for the Maneuvering Characteristics Augmentation System (MCAS), which later became a key factor in the aircraft's certification controversy. See FAA (2020) for more details.

47 Because each component can operate at its optimal speed, the number of compressor and turbine stages can be reduced, resulting in weight savings.

48 Other terms are propfan, open fan, unducted fan (UDF) or ultra-high bypass (UHB).

49 In principle, a turboprop can be seen as a type of turbofan with aerodynamically shaped, variable-pitch propeller blades instead of fan blades, without a casing or duct, and operating at an extremely high bypass ratio. Similar to the fan in a turbofan engine, propeller diameter drives efficiency. However, larger propellers increase tip speeds, which can approach or exceed the speed of sound, causing aerodynamic and noise issues. To avoid this, rotational speed must be limited, which in turn limits maximum cruise speed.

50 See IATA (2009) or Norris (2021).

51 CFM (2025).

52 IATA (2009).

53 ATI (2021).

54 For an Airbus A380 operating a flight from Paris to Singapore, approximately 119 tonnes of conventional jet fuel would be required. To supply the same amount of usable energy using current lithium-ion battery technology, a battery mass of roughly 5,272 tonnes would be needed (Eurocontrol 2023). This striking discrepancy highlights the fundamental challenges of employing batteries for large-scale aircraft propulsion.

55 Thus, issues that are parallel to those in the transition to electric vehicles for ground transportation.

56 At present, only one electric aircraft has achieved full certification: the two-seat Pipistrel Velis Electro, with a typical endurance of 50 minutes and a cruising speed of about 160 km/h (Pipistrel 2025). There are several announcements, developments and sometimes even prototypes for larger electric aircraft – but still aiming only at capacities of up to 19 passengers and short-haul ranges.

57 Green hydrogen refers to hydrogen produced from renewable energy sources. Without detailing the various production pathways, it is worth noting that currently only around 0.1% of global hydrogen production qualifies as green hydrogen (IEA 2024b).

58 See for example EASA (2025a), DLR (2025) and Proesmans and Vos (2024).

59 For the example of an Airbus A380 operating a flight from Paris to Singapore (mentioned above), approximately 730 m^3 of liquid hydrogen would be required – excluding the additional space needed for tank structures, insulation, and installation systems themselves (Eurocontrol, 2023). To put this into perspective: the total internal volume of the aircraft's three decks, including the cargo and baggage holds, is about 1,570 m^3 (Airbus 2021).

60 Here, folding wings then would also reduce the available fuel capacity within the wings, requiring more fuel to be stored in the fuselage, which in turn can reduce payload capacity.

61 Norris (2025).

62 IATA (2009) and IATA (2023).

63 Northrop Grumman (2025).

64 Furthermore, according to IATA CNG strategy, operational and infrastructure-related measures are expected to account for only about 3% of future emissions reductions in aviation.

65 According to EASA (2025a), 63% of flight emissions can be attributed to the cruise phase followed by 23% for the climb.

66 It should be noted that, particularly on long-haul flights in the Northern Hemisphere, the influence of the jet stream plays a major role in flight planning. As a result, flight paths may deviate from the great circle route – either to avoid the jet stream when flying westbound or to take advantage of it when flying eastbound in order to increase ground speed and reduce flight time.

67 According to Eurocontrol (2025b), the average en-route extension in filed flight plans within European airspace was approximately 4.5% in 2024. While air traffic control may issue in-flight shortcuts ("direct-to") during operations, the actual inefficiency still averages around 3%. Moreover, since flight dispatchers must plan fuel based on the filed route, the aircraft is loaded with fuel for the full 4.5% detour. This increases the aircraft's take-off weight and, in turn, fuel consumption – even if a shorter route is flown in practice.

68 Fichert (2023).

69 In November 2021, Airbus conducted a formation flight trial with two Airbus A350–900s from Toulouse to Montreal, maintaining a separation of approximately 1.2 NM. The test achieved fuel savings of over 5% and more than 6 tons of CO_2, with safety ensured through dedicated autopilot functions.

70 For example, most flights from Europe to North America typically depart in the morning, while services to East Asia usually leave in the afternoon. This results in recognisable and recurring flight patterns. A snapshot of major European network carriers further illustrates this: flights from their hubs to long-haul destinations such as Los Angeles and San Francisco, which follow largely similar routes, often depart within a window of just 15 minutes.

71 Of course, emissions that result from the use of office buildings or company cars are not limited to airports, but are relevant also for other entities like airlines, ANSPs and also regulatory bodies.

72 See Fraport/EnBW (2021). However, due to the specifics of the electricity market, one would have to analyse more thoroughly whether such agreements indeed increase the share of renewable electricity generation or whether they only shift emissions between different users. This aspect is not specific to airports, but to all purchase agreements on electricity from renewable sources. See for example ICCT (2020).

73 Fraport (2021).

74 Glare is a serious safety concern as it could interfere with pilots' visibility during critical flight phases. Site-specific assessments are required, focusing on panel orientation, seasonal sun angles and proximity to flight paths. For example, Amsterdam Schiphol Airport had to implement operational restrictions for specific runways caused by reflections during certain times of the day and year (Schiphol 2025).

75 Baxter and Srisaeng (2022).
76 EASA (2025a) highlights progress in forecasting capabilities and the potential for integrating contrail avoidance into existing routing operations. However, as Teoh, Schumann and Stettker (2020) point out, selecting the appropriate mitigation strategy – whether to avoid, reroute, or tolerate contrail formation – remains a complex challenge.
77 See ICAO (2021), Stankevičienė et al. (2019) for further details.
78 See Sebastian and Louis (2021) for further details and an overview about practices and challenges of waste management at airports.
79 The case of Gatwick Airport is discussed by Baxter and Srisaeng (2022) and also prominently featured by Thompson (2019) in ICAO's 2019 Environmental Report.
80 Fraport (2019b).
81 GIZ (2022).

References

Airbus (2015) 3rd Generation of Composite Materials for Airframe; World Materials Forum; https://worldmaterialsforum.com/files/downloads/3-Composites.pdf

Airbus (2021) Airbus A380 Facts and Figures; https://www.airbus.com/sites/g/files/jlcbta136/files/2021-12/EN-Airbus-A380-Facts-and-Figures-December-2021_0.pdf

Airbus (2023) The Wing Makers; https://www.airbus.com/en/newsroom/stories/2023-12-the-wing-makers

Airbus (2025) fello'fly and GEESE; https://www.airbus.com/en/innovation/future-aircraft-operations/air-traffic-management/fellofly-and-geese

Aircraft Commerce (2013) Payload and fuel burn comparison of 747–400 & 747–8F; Issue No. 89; https://www.aircraft-commerce.com/wp-content/uploads/aircraft-commerce-docs/Freight/2013/ISSUE89_FRT.pdf

Aircraft Commerce (2019) In-service performance of the PW1100G & CFM LEAP-1A; Issue No. 123; https://aircraftcommerceevents.com/article/maintenance-engineering/in-service-performance-of-the-pw1100g-cfm-leap-1a/

Aslam, M. (2025) Boeing 787 Engine Showdown: GEnx vs Trent 1000; https://www.airwaysmag.com/new-post/boeing-787-engine-genx-vs-trent-1000

ATI – Aerospace Technology Institute (2021) Realising Zero-Carbon Emission Flight; https://www.ati.org.uk/wp-content/uploads/2021/09/FZ_0_6.1-Primary-Energy-Source-Comparison-and-Selection-FINAL-230921.pdf

Avia Solutions Group (2022) The Aerodynamic Effects on Dirty Aircraft in the Quest for Greater Sustainability; https://aviasg.com/en/media/our-news/nordic-dino/the-aerodynamic-effects-on-dirty-aircraft-in-the-quest-for-greater-sustainability

Baxter, G., Srisaeng, P. (2022) Optimizing Airport Sustainable Waste Management from the Use of Waste-To-Energy Technology and Circular Economy Principles: The Case of London Gatwick Airport; in: *International Journal for Traffic and Transport Engineering*, Volume 12 (2), pp. 176–195; doi:10.7708/ijtte2022.12(2).03

Boeing (2025) Experience the 787 Dreamliner difference; https://www.boeing.com/commercial/787

Burden, P. (2025) A 1kg Weight Reduction Could Save You $4,210 in Fuel Each Year; https://atleurope.aero/a-1kg-weight-reduction-could-save-you-4210-in-fuel-each-year/

CAA – Civil Aviation Authority (2014) Managing Aviation Noise (CAP 1165); https://www.caa.co.uk/publication/download/14877

CANSO/ACI – Civil Air Navigation Services Organisation and Airports Council International (2015): Managing the Impacts of Aviation Noise – A Guide for Airport Operators and Air Navigation Service Providers; https://store.aci.aero/product/aci-and-cansos-managing-the-impacts-of-aviation-noise-a-guide-for-airport-operators-and-air-navigation-service-providers-2015/

CFM (2025) 5 Things to Know About the CFM RISE Program; https://www.cfmaeroengines.com/stories/5-things-to-know-about-rise

DLR – Deutsches Zentrum für Luft- und Raumfahrt (2025) World-first in-flight measurements of contrails from hydrogen propulsion; https://www.dlr.de/en/latest/news/2025/world-first-in-flight-measurements-of-contrails-from-hydrogen-propulsion

Dubois, T. (2020) Have The Boeing 787 And Airbus A350 Set Standard For Composites?; Aviation Week Network; https://aviationweek.com/aerospace/manufacturing-supply-chain/have-boeing-787-airbus-a350-set-standard-composites?

EASA – European Union Aviation Safety Agency (2022) Type-Certificate Data Sheet for Engine GEnx Series Engines; https://www.easa.europa.eu/en/document-library/type-certificates/engine-cs-e/easaime102-general-electric-company-genx-series

EASA – European Union Aviation Safety Agency (2025a) European Aviation Environmental Report 2025; https://www.eurocontrol.int/sites/default/files/2025-01/eurocontrol-easa-eaer-2025.pdf

EASA – European Union Aviation Safety Agency (2025b) EASA Certification Noise Levels; https://www.easa.europa.eu/en/domains/environment/easa-certification-noise-levels

East Midlands Airport (2020) Green sky thinking at East Midlands Airport; https://mediacentre.eastmidlandsairport.com/green-sky-thinking-at-east-midlands-airport/

Embraer (2021) Silence is Certifiably Golden – Embraer E-Jets; https://www.embraer.com/media-center/en/?mediatype=Insights&detail=15037-Silence-is-Certifiably-Golden-%E2%80%93-Embraer-E-Jets

Eurocontrol (2016) A-CDM Impact Assessment; https://www.eurocontrol.int/sites/default/files/2019-04/a-cdm-impact-assessment-2016.pdf

Eurocontrol (2023) The challenge of long-haul flight decarbonisation: When can cutting-edge energies and technologies make a difference?; https://www.eurocontrol.int/sites/default/files/2023-08/eurocontrol-think-paper-21-long-haul-decarb.pdf

Eurocontrol (2024) Sustainable Taxi Operations; https://www.eurocontrol.int/publication/sustainable-taxi-operations

Eurocontrol (2025a) Continuous climb and descent operations; https://www.eurocontrol.int/concept/continuous-climb-and-descent-operations

Eurocontrol (2025b) Performance Review Report; https://www.eurocontrol.int/sites/default/files/2025-03/eurocontrol-performance-review-report-2024.pdf

Eurocontrol (2025c) Vertical flight efficiency; https://ansperformance.eu/efficiency/vfe/

Eurocontrol (2025d) Total airport management; https://www.eurocontrol.int/project/total-airport-management

Evans, A., Schäfer, A. (2013) The rebound effect in the aviation sector; in: *Energy Economics*, Volume 36; pp. 158–165; doi:10.1016/j.eneco.2012.12.005

FAA – Federal Aviation Administration (2020) Summary of the FAA's Review of the Boeing 737 MAX; https://www.faa.gov/sites/faa.gov/files/2022-08/737_RTS_Summary.pdf

Fichert, F. (2023) Air navigation service provider charges; in: Efthymiou, M. (Ed.) *Air Traffic Management – Principles, Performance, Markets*; Routledge, London and New York; pp. 151–167

Fraport (2011) Connecting Sustainability. Report 2011; https://www.fraport.com/content/dam/fraport-company/documents/konzern/verantwortung/publikationen/eng/

sustainability/2011/connecting-sustainably-2011.pdf/_jcr_content/renditions/original.media_file.download_attachment.file/connecting-sustainably-2011.pdf

Fraport (2019b) Environmental Statement 2019; https://www.fraport.com/content/dam/fraport-company/documents/konzern/verantwortung/publikationen/eng/environment/environmental-statements/Umwelterkl%C3%A4rung_2019_e_online.pdf/_jcr_content/renditions/original.media_file.download_attachment.file/Umwelterkl%C3%A4rung_2019_e_online.pdf

Fraport (2022) Photovoltaic Energy Share Continues to Rise; https://www.fraport.com/en/newsroom/press-releases/2022/q4/photovoltaic-energy-share-continues-to-rise.html

Fraport/EnBW (2021) Fraport and EnBW Conclude Long-Term Power Purchase Agreement for Non-Subsidized He Dreiht Offshore Wind Farm; https://www.fraport.com/en/newsroom/press-releases/2021/q4-2021/fraport-and-enbw-conclude-long-term-power-purchase-agreement-for.html

Freeman, S., Lee, D.S., Lim, L.L., Skowron, A., De León, R.R. (2018) Trading off Aircraft Fuel Burn and NOx Emissions for Optimal Climate Policy; in: *Environmental Science and Technology*, Volume 52 (5); pp. 2498–2505; doi:10.1021/acs.est.7b05719

GIZ – Deutsche Gesellschaft für Internationale Zusammenarbeit (2022) Energy Efficiency in Airports; https://energypartnership.cn/fileadmin/china/newsroom_docs/Energy_Efficiency_in_Airports.pdf

GreenAir (2015) Major European Airbus operators take steps to avoid A320 whistle noise nuisance; https://archives.greenairnews.com/www.greenaironline.com/news1b84.html?viewStory=2116

Heathrow Airport (2025) Runway alternation programme; https://www.heathrow.com/content/dam/heathrow/web/common/documents/company/local-community/noise/operations/runway-alternation/Heathrow_Runway_Alternation_Programme_2025.pdf

Hoyas, F. (2025) Revisiting Noise Abatement Departure Procedures (NADP) Application; https://www.icao.int/sites/default/files/sp-files/environmental-protection/Documents/EnvironmentReport-2010/2025/Envreport2025_94.pdf

IATA – International Air Transport Association (2009) IATA Technology Roadmap Technical Annex; https://www.iata.org

IATA – International Air Transport Association (2018) Airport – Collaborative Decision Making (A-CDM) IATA Recommendations; https://www.iata.org/contentassets/5c1a116a6120415f87f3dadfa38859d2/iata-acdm-recommendations-v1.pdf

IATA – International Air Transport Association (2021) Resolution on the industry's commitment to reach net zero carbon emissions by 2050; https://www.iata.org/contentassets/d13875e9ed784f75bac90f000760e998/iata-agm-resolution-on-net-zero-carbon-emissions.pdf

IATA – International Air Transport Association (2023) Aircraft Technology Net Zero Roadmap; https://www.iata.org/contentassets/8d19e716636a47c184e7221c77563c93/aircraft-technology-net-zero-roadmap.pdf

IATA – International Air Transport Association (2024) Net Zero CO_2 Emissions Roadmap, Executive Summary; https://www.iata.org/contentassets/8d19e716636a47c184e7221c77563c93/executive-summary—net-zero-roadmaps.pdf

IATA – International Air Transport Association (2025a) Our Commitment to Fly Net Zero by 2050; https://www.iata.org/en/programs/sustainability/flynetzero/

IATA – International Air Transport Association (2025b) Fuel Efficiency in Aviation: Why it Matters More Than Ever; https://www.iata.org/en/publications/newsletters/iata-knowledge-hub/fuel-efficiency-in-aviation-why-it-matters-more-than-ever/

ICAO – International Civil Aviation Organization (2007) Review of Noise Abatement Procedure Research & Development and Implementation Results; https://www.icao.int/sites/default/files/sp-files/environmental-protection/Documents/ReviewNADRD.pdf

ICAO – International Civil Aviation Organization (2010) *Continuous Descent Operations (CDO) Manual*, Doc 9931 AN/476; Montreal

ICAO – International Civil Aviation Organization (2013) *Continuous Climb Operations (CCO) Manual*, Doc 9993 AN/495; Montreal

ICAO – International Civil Aviation Organization (2015) Airport Air Quality Manual; https://www.icao.int/sites/default/files/sp-files/environmental-protection/Documents/Doc%209889.SGAR.WG2.Initial%20Update.pdf

ICAO – International Civil Aviation Organization (2018) Waste Management at Airports – Eco Airport Toolkit; https://www.icao.int/sites/default/files/sp-files/environmental-protection/documents/waste_management_at_airports_booklet.pdf

ICAO – International Civil Aviation Organization (2021) Water Management at Airports; https://www.icao.int/environmental-protection/Documents/Water%20management%20at%20airports.pdf

ICAO – International Civil Aviation Organization (2022) 2022 Environmental Report: Innovation for a Green Transition; https://www.icao.int/sites/default/files/sp-files/environmental-protection/Documents/EnvironmentReport-2010/ICAO-ENV-Report-2022-F4.pdf

ICAO – International Civil Aviation Organization (2024) Presentation of 2023 Air Transport Statistics Results; https://www.icao.int/sites/default/files/environmental-protection/Documents/ARC_2023_Tables_final.pdf

ICCT – International Council on Clean Transportation (2020) Corporate Procurement of Renewable Energy: Implications and Considerations; https://www.theccc.org.uk/wp-content/uploads/2020/12/Corporate-Procurement-of-Renewable-Energy-Implications-and-Considerations-Terri-Wills.pdf

ICCT – International Council on Clean Transportation (2022) Performance Analysis of Evolutionary Hydrogen-Powered Aircraft; https://theicct.org/publication/aviation-global-evo-hydrogen-aircraft-jan22/

ICCT – International Council on Clean Transportation (2025) Understanding the greenhouse gas emissions of different SAF pathways; https://theicct.org/understanding-the-ghg-emissions-of-different-saf-pathways-sept25/

IEA – International Energy Agency (2024a) Aviation World Energy Outlook; https://https://www.iea.org/reports/world-energy-outlook-2024

IEA – International Energy Agency (2024b) Global Hydrogen Review 2024; https://www.iea.org/reports/global-hydrogen-review-2024

IPCC – Intergovernmental Panel on Climate Change (1999) Aviation and the Global Atmosphere: Demand for Air Travel; https://archive.ipcc.ch/ipccreports/sres/aviation/index.php?idp=133

Kingsley-Jones, M. (2005) Aerodynamic 'tweaks' boost Airbus A350 cruise speed; FlightGlobal; https://www.flightglobal.com/aerodynamic-tweaks-boost-airbus-a350-cruise-speed/64424.article

Koudis, G.S., Hu, S.J., Majumdar, A., Jones, R., Stettler, M.E.J. (2017) Airport emissions reductions from reduced thrust takeoff operations; in: *Transportation Research Part D:*

Transport and Environment, Volume 52, Part A; pp. 15–28; doi:10.1016/j.trd.2017.02.004

Lufthansa Cargo (2014) Cargo Climate Information; https://www.lufthansa-cargo.com/documents/20184/29985/LHC14036_Umwelteinleger_en_210x297_20140311.pdf/e2fbba44-4f5d-4bdd-a55a-c09a370d7732?version=1.0#:~:text=The%20Lufthansa%20Cargo%20freighter%20fleet,bottom%2C%20looking%20for%20unnecessary%20weight

Lufthansa Technik (2025) Cutting emissions with sharkskin technology; https://www.lufthansa-technik.com/en/aeroshark

Norris, G. (2021) Evolutionary Trail Of The Open-Fan Engine; Aviation Week Network; https://aviationweek.com/aerospace/evolutionary-trail-open-fan-engine

Norris, G. (2025) Boeing Puts X-66 On Ice But Will Continue Thin Wing Studies; *Aviation Week Network*; https://aviationweek.com/aerospace/emerging-technologies/boeing-puts-x-66-ice-will-continue-thin-wing-studies

Northrop Grumman (2025) Blended Wing Body Aircraft; https://www.northropgrumman.com/what-we-do/aircraft/blended-wing-body-aircraft

Pipistrel (2025) Velis Electro; https://www.pipistrel-aircraft.com/products/velis-electro/

Poland, M. (2020) Vertical wiper blades reduce aircraft drag, tests show; Secretary of the Air Force, Energy, Installations, and Environment; https://www.safie.hq.af.mil/News/Article-Display/Article/2153366/vertical-wiper-blades-reduce-aircraft-drag-tests-show/

Proesmans, P., Vos, R. (2024) Hydrogen, medium-range airplane design optimization for minimal global warming impact; in: *CEAS Aeronautical Journal*, Volume 15; pp. 781–806; doi:10.1007/s13272-024-00734-w

Rafi, H.A. (2024) Impact of Composite Materials on Aircraft Weight Reduction, Fuel Efficiency, and Performance in Commercial Aviation; in: *International Journal of Mechanical Engineering*, Volume 4 (1); pp. 21–35; doi:10.55640/ijme-04-01-06

Schiphol Airport (2025) Joint statement on behalf of ISMS, in which the aviation sector cooperates on safe aviation; https://news.schiphol.com/aviation-sector-takes-temporary-measure-against-glare-from-solar-panels/

Schmelcher, M., Häßy, J. (2022) Hydrogen fuel cells for aviation? A potential analysis comparing different thrust categories; in: *International Society of Air Breathing Engines*; Proceedings ISABE 2022; Ottawa, Canada; https://elib.dlr.de/188835/1/ISABE_2022_291_full.pdf

Sebastian, R.M., Louis, J. (2021) Understanding waste management at airports: A study on current practices and challenges based on literature review; in: *Renewable and Sustainable Energy Reviews*, Volume 147; 111229; doi:10.1016/j.rser.2021.111229

Stankevičienė, R., Survilė, O., Šaulys, V., Bagdžiūnaitė-Litvinaitienė, L., Litvinaitis, A. (2019) The Treatment and Handling Systems of De/Anti-Icing Contaminants which Generated and Discharged into Surface Runoff from Airports Territories; in: *Journal of Water Security*, Volume 5; jws2019005; doi:10.15544/jws.2019.005

Steinegger, R. (2017) Fuel Economy as Function of Weight and Distance; Zürcher Fachhochschule; https://digitalcollection.zhaw.ch/server/api/core/bitstreams/e39647c4-dd8b-4461-9f2d-afeb74574209/content

Sunaryo, E.R., Zuniga, C. (2025) The future is here: Airports enter the age of automation; https://www.internationalairportreview.com/article/260776/the-future-is-here-airports-enter-the-age-of-automation/

Teoh, R., Schumann, U., Stettler, M.E.J. (2020) Beyond Contrail Avoidance: Efficacy of Flight Altitude Changes to Minimise Contrail Climate Forcing; in: *Aerospace*, Volume 7 (9); 121; doi:10.3390/aerospace7090121

Thompson, R. (2019) Gatwick's Circular System for Becoming a Zero Waste Airport; in: *ICAO 2019 Environmental Report*; pp. 300–305; https://www.icao.int/sites/default/files/sp-files/environmental-protection/Documents/EnvironmentReport-2010/ICAO-ENV-Report2019-F1-WEB-1.pdf

Wallington, T.J., Woody, M., Lewis, G.M., Keoleian, G.A., Adler, E.J., Martins, J.R.R.A., Collette, M.D. (2025) Hydrogen as a sustainable transportation fuel; in: *Renewable and Sustainable Energy Reviews,* Volume 217; 115725; doi:10.1016/j.rser.2025.115725

Zurich Airport (2017) Taxi-Emissions at Zurich Airport; https://www.flughafen-zuerich.ch/-/jssmedia/airport/portal/dokumente/das-unternehmen/politics-and-responsibility/environmental-protection/technische-berichte/2017_taxi_study_zurichairport.pdf

Zurich Airport (2018) Aircraft Ground Energy Systems; https://media.flughafen-zuerich.ch/-/jssmedia/airport/portal/dokumente/das-unternehmen/politics-and-responsibility/environmental-protection/technische-berichte/2018_zrh_aircraft-ground-energy-system.pdf

Zurich Airport (2023) Underground subglacial channel to enable emission-free heating and cooling at Zurich Airport; https://newsroom.flughafen-zuerich.ch/en/underground-subglacial-groove-to-enable-emission-free-heating-and-cooling-at-zurich-airport/

7 Sustainability Reporting in Aviation

Introduction

In recent years, Sustainability Reporting has evolved into a central instrument for ensuring transparency and accountability in corporate practice. As environmental, social and governance (ESG) issues gain increasing relevance for regulators, investors and the public, structured sustainability disclosures have become an integral part of corporate strategy. The aviation sector – highly visible, emissions-intensive and globally interconnected – is particularly affected by these developments. This chapter provides a comprehensive overview of the evolution, structure and regulatory environment of sustainability reporting, with a particular focus on its implications for the aviation industry.

Background for Sustainability Reporting

Sustainability Reporting is an instrument of non-financial disclosure that allows companies to document and communicate their progress and performance on central sustainability issues, typically structured around Environmental, Social and Governance (ESG) factors. While many reporting practices are still voluntary, an increasing number of jurisdictions have introduced mandatory disclosure requirements, especially for large companies and listed entities. As a result, Sustainability Reports are now published both on a voluntary basis and to comply with legal obligations. According to KPMG (2024a), 96% of the world's 250 largest companies (G250) publish Sustainability or ESG reports and 79% of the 5,800 largest companies surveyed worldwide (N100) do the same.[1]

Sustainability Reporting creates value not only in meeting legal obligations but also in the context of voluntary corporate transparency. By making a company's sustainability performance transparent to stakeholders, it enables organisations to identify and mitigate risks, improve operational and corporate efficiency and strengthen trust with investors and employees. Enhanced transparency can also improve access to capital, as several studies show that visible ESG disclosure reduces information asymmetry and lowers the cost of capital through higher valuation and lower perceived risk.[2] In general, Sustainability Reporting is increasingly influencing decision-making in the financial

DOI: 10.4324/9781003608172-7

community, as investors, asset managers and rating agencies integrate ESG information into their risk assessments and capital allocation decisions.[3]

Sustainability Reporting in Aviation

While the general motive for Sustainability Reporting applies to aviation as it does to other industries, it is of particular relevance due to the characteristics of the sector. Industry organisations such as IATA (2024a), Airlines for Europe (A4E 2025) and ACI (2025) acknowledge that, as a globally visible and "hard-to-abate" sector, aviation is under close public and regulatory scrutiny due to its considerable environmental footprint. The urgency to act is driven not only by aviation's role in global warming but also from local impacts such as noise and air quality (see Chapter 4). Therefore, transparent and credible reporting is essential to show real mitigation efforts and build stakeholder trust.[4]

However, the industry also faces specific and complex reporting challenges. Like other industries, aviation is confronted with a fragmented landscape of evolving regulations and standards.[5] However, this challenge is amplified by the global nature of the sector. E.g., airlines operating across multiple jurisdictions must comply with a patchwork of national and international requirements, which often have diverging scopes and methodologies. This complexity is further increased by the inconsistent use of metrics. While regulators often demand absolute emissions data, many operators still favour relative intensity indicators to show operational efficiency.[6] In addition, airlines often use different calculation methods, which further limits consistency and comparability.[7] As a result, comparisons remain challenging not only between individual companies but also across different segments of the aviation value chain.

Content of Sustainability Reporting

In general, a Sustainability Report is a form of non-financial disclosure that complements traditional financial reporting. The purpose is to create transparency about sustainability-related risks, inform stakeholders (e.g., investors) and report on strategic sustainability goals and their progress. This often also meets statutory reporting obligations (e.g., CSRD in the EU). Reporting frameworks and standards each provide the structured guidance for this process. They define which topics are key, which metrics should be used and what the criteria for measuring and reporting the results should be in order to ensure transparency and comparability.

While the specifics can vary, the core components of a comprehensive ESG report typically include a mix of quantitative indicators[8] and qualitative information:[9]

- Environmental Information: This is often the most data-intensive part. Key indicators include greenhouse gas (GHG) emissions as well as emissions of other air pollutants such as nitrogen oxides (NO_x), sulphur

oxides (SO_x), particulate matter (PM), but also noise. Many reports also provide data on energy consumption (e.g., total energy use and share of renewable energy), water withdrawal and discharge, waste generation and recycling rates and in some cases biodiversity aspects, such as land use or habitat conservation efforts.
- Social Information: Social indicators address an organisation's impact on employees, customers and communities. Common workforce metrics include health and safety performance (e.g., work-related injury rate resulting in lost time), diversity and inclusion (e.g., gender balance in leadership positions, representation of minorities) and training and development (e.g., average training hours per employee). Beyond the workforce, the reports often refer to compliance with labour rights in the supply chain, customer health and safety (e.g., product safety recalls or customer complaint rates) and contributions to community development, such as education or social investment programmes. Some companies also report on employee satisfaction or turnover as an indicator of workplace quality.
- Governance Information: Governance indicators focus on how sustainability commitments are integrated into corporate decision-making and oversight. Typical disclosures include board-level responsibility for ESG topics, the existence of a dedicated sustainability or ethics committee and the proportion of board members with ESG-related expertise. Ethical business conduct is another key topic, often reported through the number of anti-corruption training hours, whistleblower reports, or confirmed cases of fraud and corruption. A growing number of companies also disclose whether ESG performance is linked to executive compensation, for example by tying bonuses to emission reduction targets or diversity goals.

An important aspect of Sustainability Reporting is the distinction between general and sector-specific metrics. While some indicators like GHG emissions are universal (or cross-sectional), many frameworks now require industry-specific (or sector-specific) disclosures in order to reflect the individual impacts of different sectors. This enables a more meaningful comparison between peers.[10]

Recently, there has been a trend toward incorporating forward-looking information. Increasingly, organisations are expected to disclose risk assessments, transition plans and long- and mid-term emission reduction targets to enable stakeholders to evaluate the firms' resilience to climate change and other sustainability challenges.[11]

Selected Leading Sustainability Reporting Frameworks

Although still technically voluntary in many jurisdictions, the following initiatives have become the most influential global frameworks for Sustainability

Reporting. They are characterised by broad international recognition, regulatory relevance and widespread adoption by companies and investors. Together, they form the de facto global baseline for ESG reporting, focusing on initiatives with global relevance and cross-sector applicability.

- Global Reporting Initiative (GRI): The GRI was founded in 1997 as an international, independent organisation and was the first to establish globally recognised standards for Sustainability Reporting.[12] Initially focused on environmental accountability, its scope was later expanded to include ESG issues. In 2016, the GRI transitioned to the modular GRI Standards, applicable to organisations of all sizes and sectors and structured into universal, sector-specific and topic-specific standards covering a broad range of ESG aspects, including widely used indicators such as GRI 302 (Energy) and GRI 305 (Emissions). Today, the "GRI Standards" are the most widely applied Sustainability Reporting framework worldwide.[13]
- Sustainability Accounting Standards Board (SASB): The SASB, established in 2011 as a non-profit organisation, was designed to provide companies and investors with a consistent framework for the disclosure of sustainability-related issues with financial relevance.[14] The main objective is to standardise the publication of ESG topics that are relevant to investors in order to improve the comparability, consistency and reliability of ESG data in investment decisions. The "SASB Standards" cover 77 industries, providing sector-specific metrics tailored to the ESG topics most relevant to financial performance, thereby enabling investors to more effectively integrate ESG considerations into risk assessment and corporate valuation processes.[15]
- Task Force on Climate-related Financial Disclosures (TCFD): The TCFD is a global framework established in 2015 by the Financial Stability Board to support consistent and decision-relevant disclosure of climate-related financial risks and opportunities.[16] Aimed primarily at investors, lenders and insurers, the TCFD recommendations are structured around four core pillars (Governance, Strategy, Risk Management and Metrics & Targets).[17]
- International Sustainability Standards Board (ISSB): The ISSB was established in 2021 by the International Financial Reporting Standards (IFRS)[18] Foundation to provide investors and other capital market participants with consistent and comparable information on sustainability-related risks and opportunities.[19] Building on existing investor-focused initiatives like the SASB Standards and TCFD recommendations, the ISSB framework combines sector-specific, financially material indicators with forward-looking governance and risk management elements.[20] The "ISSB Standards" are designed to serve as a global baseline for Sustainability Reporting, allowing jurisdictions to adopt them directly or integrate them into national regulatory frameworks.[21]

- Carbon Disclosure Project (CDP): The Carbon Disclosure Project, now operating simply as CDP, was founded in 2000 as a not-for-profit organisation to encourage companies and cities to disclose their environmental impacts.[22] It has become one of the largest global disclosure systems for environmental performance, focusing on climate change, water security and deforestation. Participating companies and cities submit responses to standardised questionnaires, and CDP assigns scores ranging from A to D, based on disclosure quality, risk management and the adoption of environmental best practices. These scores are widely used by investors to benchmark corporate environmental leadership and evaluate environmental performance.[23]
- Greenhouse Gas (GHG) Protocol: The GHG Protocol provides the methodological foundation for many major Sustainability Reporting frameworks, including GRI, TCFD, ISSB and CDP, by establishing globally standardised rules for calculating and categorising greenhouse gas emissions.[24] Its primary objective is to support consistent measurement and management of GHG emissions. The Protocol distinguishes between Scope 1 emissions from owned or controlled sources, Scope 2 emissions from purchased energy and Scope 3 emissions covering all other indirect emissions across the value chain (see also Chapters 4 and 8).[25] Because of its clear methodology and broad international adoption, the GHG Protocol is widely used as the standard framework for emissions accounting; for example, already in 2016, 92% of Fortune 500 companies applied it when responding to CDP.[26]

Additional International Sustainability Reporting Frameworks

In addition to the frameworks outlined above, there is a wide range of further initiatives and standards that play a significant role in Sustainability Reporting practices. While not all of them are used as primary reporting frameworks, they provide valuable guidance, complement existing standards or support the implementation of broader sustainability goals. The following examples illustrate some of the most widely recognised among these initiatives:

- Climate Disclosure Standards Board (CDSB): The CDSB is an international consortium of businesses and environmental non-governmental organisations (NGOs) that provides a framework for reporting environmental and climate-related information within mainstream financial reports. Initially focused on climate change and GHG emissions, the framework was later expanded to include broader aspects.[27]
- United Nations Sustainable Development Goals and Global Compact: The UN Sustainable Development Goals (SDGs) provide a global reference framework that organisations can use to align their strategies and reporting to demonstrate their contribution to sustainable development.[28] One of the key implementation platforms is the UN Global Compact (UNGC), a voluntary initiative that encourages companies to

align their operations with ten universally accepted principles to deliver SDGs through responsible business practices.[29]

- Principles for Responsible Investment (PRI): The UN-supported PRI initiative provides a voluntary framework for institutional investors to incorporate ESG factors into investment and ownership decisions, based on six guiding principles.[30]
- ISO 26000 (Social Responsibility Guidance Standard): ISO 26000 offers guidance on how organisations can operate in a socially responsible way. It covers seven core subjects, including environmental issues and supports transparent communication of social responsibility efforts, but it is not a certifiable standard.[31]

From Voluntary to Mandatory Sustainability Reporting

Although many Sustainability Reporting initiatives are formally voluntary, they have become de facto mandatory due to investor pressure, ESG-based financial ratings and general market expectations.[32] Large asset owners, rating agencies and stock exchanges rely heavily on data structured according to established frameworks to assess risk, allocate capital and benchmark corporate performance with non-reporting companies facing lower ESG scores, potential reputational disadvantages and higher financing costs.[33] As a consequence, frameworks such as the GRI Standards dominate cross-sectoral Sustainability Reporting, complemented by investor-focused initiatives such as SASB and TCFD. Disclosure platforms like the CDP have also become market expectations,[34] while the ISSB Standards are already moving towards a quasi-mandatory role as regulators and stock exchanges begin to integrate them into their listing rules.[35]

Building on this market-driven adoption, an increasing number of jurisdictions is now embedding leading voluntary frameworks into legally binding sustainability disclosure requirements.[36] The GRI (2025) Standards are referenced in more than 160 governmental and regulatory policies across over 60 countries, and the TCFD recommendations have evolved into a global regulatory reference for climate-related financial disclosures in major markets including the United Kingdom, Japan and New Zealand with further integration underway in Australia, Canada and the USA.[37]

These developments illustrate how voluntarily implemented guidelines are gradually transforming into formal legal obligations. The most prominent example is the European Union's Corporate Sustainability Reporting Directive (CSRD), which has established a comprehensive, legally binding framework and could be regarded as a global benchmark for Sustainability Reporting.

EU's Corporate Sustainability Reporting Directive

The Corporate Sustainability Reporting Directive (CSRD, EU 2022/2464) establishes a mandatory and standardised legal framework for Sustainability

Reporting in the European Union.[38] Adopted in 2022, it significantly expands the scope and detail of reporting requirements compared to its predecessor, the Non-Financial Reporting Directive (NFRD), with first reports due in 2025 for large companies.

In 2025, the European Commission (EC 2025h) adopted adjustments to the CSRD implementation timeline to avoid overburdening businesses and to preserve the EU's economic competitiveness. Compared to the original CSRD schedule, these adjustments include a reduction in the number of companies[39] subject to reporting obligations, a two-year postponement for companies not previously covered by the NFRD and revisions aimed at reducing the amount and complexity of data to be disclosed.

A key feature of the CSRD is the principle of "double materiality", requiring companies to not only disclose how sustainability issues affect their financial performance (outside-in perspective) but also how their activities impact the environment and society (inside-out perspective). This approach ensures that reporting serves not only investors but also a broader range of stakeholders.[40]

To promote consistency and comparability, the CSRD introduces the European Sustainability Reporting Standards (ESRS), developed by the European Financial Reporting Advisory Group (EFRAG)[41] in alignment with international frameworks. The ESRS are closely coordinated with the ISSB Standards, build on TCFD for climate-risk disclosures and refer to the GHG Protocol for Scope 1, Scope 2 and Scope 3 emissions. Many ESG topics are also consistent with GRI Standards, which supports interoperability and reducing the reporting burden for multinational companies.[42]

The CSRD also strengthens governance and assurance requirements by requiring that sustainability information be integrated into the management report, subjected to external assurance (initially in the form of "limited assurance") and electronically reported according to given reporting format standards to enable automated data processing and comparability across sectors.[43]

By setting these stringent requirements, the CSRD elevates Sustainability Reporting to a level comparable with financial reporting. Its influence extends far beyond the EU, as many non-EU companies with EU parent or subsidiary companies or with significant operations (revenue) in Europe will also have to comply, and other jurisdictions are closely observing the EU's approach as a potential model for future regulation.[44]

Evolving Global Sustainability Reporting Landscape

The previous section has shown that globally recognised Sustainability Reporting initiatives from early voluntary frameworks such as the GRI Standards have evolved to increasingly quasi-mandatory investor-driven practices (e.g., SASB, TCFD, CDP, ISSB). Many are now explicitly referenced by regulators and are gradually transitioning into mandatory disclosure requirements, with the EU's CSRD as the most prominent example.

Alignment efforts, e.g., through the ISSB, and an increasing number of national disclosure rules aim to harmonise standards but also create uncertainty and complexity, as companies must continuously adapt to changing requirements. And the reporting landscape remains heterogeneous and fragmented. Frameworks differ in purpose and scope from investor-focused standards (SASB, TCFD, ISSB) to multi-stakeholder initiatives (GRI) and environmental disclosure platforms (CDP). Some address the full ESG spectrum, others focus on specific domains such as GHG emissions accounting (GHG Protocol). Indicators vary between cross-sectoral metrics and sector-specific measures, and initiatives differ in nature from voluntary disclosure platforms to formal standard-setting bodies.

Implications for the Aviation Sector

Aviation companies are subject to all these general challenges, along with additional sector-specific complexities. Like other industries, airlines and airports must align their reporting with, where applicable, legally binding frameworks such as the CSRD and the GRI Standards, SASB, TCFD, CDP, ISSB, to meet investor expectations and maintain access to capital markets. However, the characteristics of the aviation sector make compliance even more challenging.[45] Its global nature means that aviation companies might have to respond to a mix of voluntary initiatives, investor-driven disclosure requirements and jurisdiction-specific regulations,[46] often with differing methodologies and metrics.

This inconsistency in transparency and reporting practices has been highlighted not only by industry stakeholders but also by independent academic research, which increasingly calls for harmonisation and standardisation in Sustainability Reporting.[47]

The lack of consistent, industry-specific Sustainability Reporting guidelines also complicates the integration of aviation-relevant indicators into predominantly cross-sectoral frameworks. Thus, the aviation sector must adapt to changing global frameworks while at the same time integrating aviation-specific indicators that are not yet fully standardised across industries.

Sector-Specific Initiatives

To better address these challenges, the aviation industry has developed sector-specific initiatives that complement the general Sustainability Reporting framework and reflect its unique operational and environmental profile. These initiatives support airlines and airports in fulfilling regulatory obligations while integrating aviation-specific indicators. Developed by industry associations and international organisations, they provide standardised methodologies, comparable metrics and verification processes tailored to aviation and are increasingly recognised by regulators and investors.

The most relevant sector-specific initiatives include:

- IATA ESG Metrics: Introduced by IATA, these metrics offer airlines a harmonised set of environmental, social and governance indicators. They combine absolute and intensity-based emissions data with social and governance KPIs to enhance comparability and alignment with regulatory and investor expectations.
- ACI Initiatives including Airport Carbon Accreditation: Airports Council International (ACI) is developing guidance to improve and harmonise ESG reporting practices at airports. In North America, ACI-NA has already published dedicated ESG reporting guidelines. A key initiative with a focus on carbon management is the Airport Carbon Accreditation (ACA) programme, which has become the globally recognised standard for measuring, reducing and verifying airport-related CO_2 emissions through a tiered certification system.
- Reporting under CORSIA and the EU Emissions Trading System (EU ETS): Both the ICAO-led CORSIA scheme and the EU ETS rely on mandatory monitoring, reporting and verification (MRV) of CO_2 emissions. While primarily designed for offsetting purposes, the resulting data could also be used in structured Sustainability Reporting.
- EU MRV for non-CO_2 effects and NEATS: Beginning in 2025, the EU has started to implement a dedicated MRV framework to quantify non-CO_2 climate effects from aviation, such as contrails and NO_x emissions. A central element is the "Non-CO_2 Aviation Effects Tracking System" (NEATS), which uses flight trajectory and atmospheric data to calculate climate impacts in CO_2-equivalent terms.

IATA ESG Metrics

The IATA ESG Metrics were introduced in 2022 as part of the Airline Sustainability Reporting Handbook (ASRH) to provide airlines a structured and globally consistent approach to ESG reporting.[48] They offer a consistent set of quantitative airline-specific indicators while ensuring strong alignment with established reporting standards.

The development of the IATA ESG Metrics was based on a broad review of existing airline reports, ESG indicators and existing and proposed ESG regulations and frameworks[49] to ensure consistency with industry and investor expectations.

The resulting guidance framework specifies ESG indicators, divided into "Core" and so-called "Core+" metrics.[50] Environmental metrics include absolute emissions (Scopes 1, 2 and selected Scope 3) as well as intensity indicators like CO_2 per RPK or RTK. Additional KPIs relate to fuel efficiency, SAF usage and waste per passenger. Social indicators cover employee diversity, training and safety performance (e.g., Lost-Time Injury Rate). Governance metrics focus on board-level oversight of sustainability risks and the presence of dedicated ESG management structures.

To provide a more detailed overview of the structure and thematic coverage of the framework, Table 7.1 presents IATA's ESG Metrics, distinguishing between "Core" and "Core+" indicators across the environmental, social and governance dimensions. Further details on the individual metrics and their alignment with existing reporting standards can be found in IATA's "Beginner's Guide to Airline Sustainability Reporting".[51]

ACI Initiatives

Similar to IATA's efforts to harmonise ESG reporting for airlines, Airports Council International (ACI) is working to improve and standardise Sustainability Reporting across the airport sector. At the global level, ACI is currently developing a dedicated ESG reporting framework for airports.[52] This initiative builds on the ACI ESG Management Best Practice[53] guidance and contributions from various regional ACI offices.

ACI North America has already developed a more detailed approach.[54] Recognising that Europe is ahead of North America in terms of ESG regulations, ACI-NA has developed an ESG framework that is based on leading global standards and aligned with the information needs of capital markets.[55]

Table 7.2 provides an overview of the framework, which covers disclosures across the environmental, social and governance pillars, divided

Table 7.1 Overview of IATA ESG metrics.

	Core	*Core+*
Environmental metrics	• Fuel consumption • Scope 1 & 2 CO_2 emissions • NO_x emissions (LTO) • Aircraft noise • Energy consumption	• Scope 3 CO_2 emissions • Carbon offsets (voluntary/mandatory) • Waste volumes and recyclables volumes/share of total waste • Water consumption • Single-use plastics purchased
Social metrics	• Community engagement • Work-related injuries and fatalities • Employee training	• Work-related ill-health and fatalities (ill-health) • Employee satisfaction • Aviation accidents • Child or forced labour risks • Pay gap women to men
Governance metrics	• Diversity • Corruption cases	• Environmental and social screening of suppliers • Collective bargaining coverage • Monetary impact and legal cases related to anti-competitive behaviour • Safety regulation enforcement

Table 7.2. Overview of ESG metrics in the ACI framework.

	Recommended	*Optional*
Environmental disclosures	• Energy consumption and intensity • GHG emissions and intensity (Scope 1 and 2) • Climate risk exposure and adaptation • Environmental commitments and progress • Environmental non-compliance • Waste management • Water management	• Energy consumption by source • Energy management • GHG from other sources (Scope 3) • GHG reduced from base year • Decarbonisation strategy and initiatives • Total waste generated • Waste diversion and disposal • Water withdrawal by source • Biodiversity management • Noise complaints and noise actions taken
Social disclosures	• Diversity, equity and inclusion (DEI) – workforce ethnicity and gender • Business supplier diversity • Concessionaire diversity • Workplace/employee safety	• DEI – by age bracket and by new hires • Labour management • Economic impact of airport on region • Economic impact – jobs created • Customer satisfaction • Construction/safety accidents • Employee participation in wellness programs • Health and safety practices • Employee engagement • Employee training • Environmental justice • Community engagement • Human trafficking initiatives

	Recommended	*Optional*
Governance disclosures	• Governance/organisational structure • Board of directors' background/diversity • Executive management background/diversity • Approach to risk management • Cyber security – general discussion • Reporting and transparency • Ethics and compliance	• Leadership system • Cyber security – specific discussion • Financial risks

into recommended and optional metrics. Further details on the individual indicators including their mapping to GRI Standards are available in ACI's "Introduction to Performance and Risk Management Metrics for North American Airports".[56]

While the development of the ACI ESG framework is still relatively new, the ACA programme has already become the globally recognised initiative for carbon management, reporting and certification in the airport sector. ACA (2025) is the only internationally endorsed carbon certification scheme specifically designed for airports and is supported by institutions such as ICAO, Eurocontrol, the FAA and others.

The programme provides a structured pathway to measure, reduce and ultimately eliminate airport-related carbon emissions. Its framework consists of seven certification levels, each representing a progressively more ambitious commitment:

- Level 1 Mapping: Establishment of a verified carbon footprint report for Scope 1 and 2 emissions and a formal commitment to CO_2 reduction,
- Level 2 Reduction: Implementation of a carbon management plan and achievement of annual CO_2 reductions relative to a three-year rolling average,
- Level 3 Optimisation: Expansion of the carbon footprint to include selected Scope 3 emissions (e.g., ground access, APU use), along with a Stakeholder Engagement Plan,
- Level 3+ Neutrality: Additional offsetting of all residual Scope 1 and 2 emissions,
- Level 4 Transformation: Introduction of long-term absolute reduction targets including a Stakeholder Partnership Plan,
- Level 4+ Transition: Additional offsetting of remaining direct emissions,
- Level 5: More than 90% absolute CO_2 reductions in Scope 1 and 2 emissions and a commitment to Net Zero for Scope 3 by 2050 or earlier.

ACA participation is subject to independent third-party verification. As of mid-2025, 664 airports in 97 countries representing more than 55% of global passenger traffic were accredited. Among them, 100 airports achieved Levels 4 or 4+ and 24 reached Level 5 Net Zero. In the reporting year 2023/24 alone, accredited airports collectively reduced 1.04 million tonnes of CO_2 from Scope 1 and 2 emissions that corresponds to a 14.8% reduction compared to the three-year rolling average.[57]

Reporting under CORSIA and the EU ETS

CORSIA (Carbon Offsetting and Reduction Scheme for International Aviation) was adopted by ICAO in 2016 and is the first global, sector-specific climate policy instrument in aviation and relies on an MRV system.[58] While CORSIA's main objective is to offset CO_2 emissions from international aviation (further details in Chapter 5), its credibility and effectiveness rely on the quality of the reporting scheme. According to IATA (2025), all airline operators with annual CO_2 emissions exceeding 10,000 tonnes have been required since 1 January 2019 to report their emissions from international flights on an annual basis. Operators must monitor fuel use and calculate (or use the CORSIA CO_2 Estimation and Reporting Tool to estimate) their emissions. All reported data must be externally verified by accredited third-party verifiers prior to submission to the respective ICAO Member State, which in turn transmits the aggregated emissions data to ICAO for publication by operator and by state pair. As of mid-2025, 129 states voluntarily participated in CORSIA, covering 99% of international CO_2 emissions from the 2023 baseline year.[59]

Similar principles as in CORSIA apply to the cross-sectoral EU ETS, which also requires MRV as a prerequisite for market-based compliance rather than as an end in itself.[60] All airlines operating flights within the European Economic Area (EEA) are required to monitor, verify and annually report their CO_2 emissions under an approved monitoring plan. All emission reports must follow GHG Protocol Scope 1 principles and be verified by an accredited third-party verifier and reports must be submitted by 31 March of the following calendar year.[61]

EU MRV for non-CO_2 Effects and NEATS

So far, the EU ETS has covered only CO_2 emissions from aviation. Given the significant contribution of non-CO_2 effects to the aviation sector's overall climate impact (see Chapter 4), the European Commission (2025f) has introduced an MRV system for non-CO_2 aviation effects, effective from 1 January 2025. The system aims to quantify climate impacts per flight in terms of CO_2 equivalents (CO_2e), using modelling approaches that combine flight data, aircraft and fuel characteristics, operational performance parameters and meteorological inputs.

The proposed framework represents the first initiative to operationalise non-CO_2 MRV in aviation and is closely aligned with the existing EU ETS structure. Reporting must be carried out annually, with verification by accredited third parties and submission by 31 March. A progress report on the system's performance is expected by the end of 2027 and may inform future legislative proposals.

To support implementation, an automated IT platform called NEATS (Non-CO_2 Aviation Effects Tracking System) is developed that calculates non-CO_2 effects (such as contrail formation and nitrogen oxide (NO_x) emissions) and their radiative forcing impact.[62] The system integrates aircraft, fuel, flight trajectory and weather data with open-source climate response models.

The MRV system is based on both Primary data (supplied by aircraft operators) and Secondary data (automatically sourced by NEATS):

- Primary data includes operator-monitored values such as actual 4D flight trajectories, aircraft mass, engine identifiers, fuel flow and detailed fuel properties (like aromatics content, hydrogen-to-carbon ratio, net calorific value),
- Secondary data is automatically sourced by NEATS and includes flight plans, trajectory data (for example from Eurocontrol), weather forecasts/ models and conservative default values for aircraft and fuel parameters (if operators do not provide this information).[63]

NEATS processes these inputs using a chain of scientific modules covering fuel consumption, contrail formation and climate impact modelling. At the end of each reporting year, the platform automatically generates a digital report for each flight. While NEATS is the only reporting tool available in 2025, its use is not mandatory in the long term. From 2026 onwards, aircraft operators may choose to use Commission-approved third-party IT tools, provided these meet official validation requirements. The selected tool must be specified in the operator's monitoring plan.

This new MRV obligation applies to all aircraft equipped with jet engines (turboprops are excluded). The geographical coverage will be implemented in phases. Initially, the system applies to flights within the EEA. Then, selected extra-EEA routes until finally a full scope is reached, encompassing all outbound and inbound flights between the EEA and third countries.

Although innovative in its approach, the new non-CO_2 MRV scheme has raised some concerns from industry stakeholders and aviation associations. These relate primarily to methodological uncertainties, the use of conservative default values that could lead to an overestimation of non-CO_2 effects and potential operational trade-offs like the risk of increased CO_2 emissions because of contrail-avoidance routing.[64]

The European Commission has acknowledged these concerns and plans to refine the system once empirical validation data becomes available and a review of the MRV framework is planned for late 2027. Nevertheless, the new framework represents an important step forward in addressing aviation's full climate footprint. By providing a consistent basis for quantifying non-CO_2 effects and incorporating them into structured reporting, the system creates a foundation for future initiatives to mitigate the climate impact from aviation.

Conclusion and Outlook

As in many other industries, Sustainability Reporting in aviation has evolved from voluntary disclosures to a requirement driven by regulation, investor demands and public expectations. However, it still faces several challenges. Although initiatives such as the ISSB aim to align with established frameworks like GRI and TCFD the global reporting landscape remains fragmented with differences in definitions, scope and metrics between various sectors and jurisdictions.

In this context, aviation companies must comply with a wide range of cross-industry reporting requirements on sustainability. At the same time, there are several industry-specific initiatives on Sustainability Reporting with the goal of better meeting both regulatory requirements and market expectations. Although these aviation-focused efforts may help to close existing gaps, actual comparability between industry sectors remains limited. Maintaining consistency while accommodating sector-specific requirements will remain a challenge in the further development of Sustainability Reporting.

In the future, the scope of reporting requirements is expected to expand. The EU's integration of non-CO_2 effects into mandatory MRV frameworks already reflects a shift towards a more comprehensive coverage of aviation's climate footprint. As a result, Sustainability Reporting will increasingly evolve beyond information on CO_2 emissions.

The digitisation of reporting is also gaining importance. As transparency requirements become increasingly detailed and complex, aviation firms are turning to automated data collection and processing tools to efficiently generate ESG information. Regulatory frameworks such as the EU's NEATS platform demonstrate how digital MRV systems can support compliance while enabling structured, machine-readable data formats. Another trend is the growing importance of verification and assurance. While schemes like CORSIA and ACA already require external verification of emissions data, the CSRD introduces formal audit obligations, marking a shift toward broader assurance requirements.

Finally, it is important to note that current reporting practices remain primarily focused on the environmental dimension. While environmental metrics (particularly greenhouse gas emissions) remain the focal point of regulation and stakeholder interest, future developments must similarly address social and governance-related aspects. This is the only way to ensure that ESG reporting evolves into a truly holistic model.

Key Takeaways

- Sustainability Reporting has evolved from voluntary practice into a core strategic and regulatory requirement driven by investors, regulators and public scrutiny.
- Aviation faces particularly high reporting challenges due to its global footprint, emissions intensity and hard-to-abate characteristics.
- The Sustainability Reporting landscape remains fragmented, creating significant complexity for aviation companies operating across multiple jurisdictions. Also, inconsistent metrics and methodologies limit transparency and comparability.
- Environmental reporting dominates current practice, but comprehensive sustainability reports combine many ESG indicators (quantitative and qualitative disclosures).

- Global frameworks such as GRI, SASB, TCFD, ISSB, CDP and the GHG Protocol form the de facto baseline for ESG reporting.
- Many formally voluntary frameworks have become quasi-mandatory due to capital market expectations and ESG-based risk assessments. Mandatory regimes, most notably the EU's CSRD, elevate Sustainability Reporting to a level comparable with financial reporting.
- Sector-specific initiatives (e.g. IATA ESG Metrics, ACI frameworks, ACA, CORSIA, EU ETS and non-CO_2 MRV/NEATS) are essential to reflect aviation's specific impacts and enable meaningful comparability.

Notes

1 G250 and N100 refer to the research samples defined by KPMG for their survey. The G250 comprises the top 250 companies from the Fortune Global 500 ranking, while the N100 consists of the top 100 companies by revenue in each of the 58 countries surveyed (KPMG 2024a).
2 See for example Economidou et al. (2023), Fenili and Raimondo (2021) and Reber, Gold and Gold (2022).
3 See for example KPMG (2024a) and Joseph (2024). According to PwC's Global Investor Survey (2024) "71% of investors agree that companies should incorporate ESG/sustainability directly into their corporate strategy".
4 However, according to Johansson (2022), only about one third of the major passenger airlines worldwide published a dedicated and publicly available comprehensive sustainability report in 2019 (with The Americas having the highest share of airlines publishing sustainability reports (51%) and Africa & Middle East having the lowest (8%)). A similar picture emerges in the airport sector: only 8 out of 30 major U.S. hubs issued a stand-alone sustainability or ESG report (Hubbard 2025). It should be noted, however, that these figures refer specifically to dedicated reports; some sustainability-related information may still be disclosed through other channels.
5 For example, Carrots & Sticks (2024) reports almost 2,700 distinct ESG and sustainability policies worldwide, representing a 9% increase compared to 2023. Carrots & Sticks (2020) noted that in 2016 there were only about 400 reporting provisions, illustrating the rapid growth in Sustainability Reporting requirements.
6 Absolute emission data, such as total tonnes of CO_2, is required under regulatory frameworks like the CSRD and CORSIA. Relative intensity indicators commonly used in aviation include grams of CO_2 per revenue tonne-kilometre (RTK) or per available tonne-kilometre (ATK).
7 IATA (2024a) and SFC (2025).
8 Beyond the selection of indicators themselves, the way in which data is presented can significantly influence stakeholder perception, regardless of whether the topic is financial or non-financial. Decisions about whether to report absolute values (e.g., total CO_2 emissions) or relative metrics (e.g., CO_2 per RPK) are not purely technical but often strategic in nature. Intensity metrics can illustrate efficiency gains, but they may obscure absolute growth in environmental or social impacts. Likewise, the choice of graphical formats, baseline years, or data aggregation can influence how issues are framed, either highlighting or downplaying developments. These are not challenges unique to Sustainability Reporting but reflect broader issues in corporate communication, where methodological transparency and consistency are key to credible reporting. A deeper examination of these representational dynamics, however, lies beyond the scope of this chapter.

9 For a comprehensive overview of ESG disclosure topics, including detailed indicators and recent developments, see the Global Reporting Initiative (GRI 2025). As one of the first and most widely adopted Sustainability Reporting frameworks worldwide, the GRI provides structured guidance on a broad range of environmental, social and governance issues and continues to play a key role in shaping global reporting practices.

10 For example, universal indicators include metrics like total CO_2 emissions or gender diversity, which apply across all industries. In contrast, sector-specific indicators address impacts unique to specific industries. As examples, fuel efficiency or noise emissions would be relevant indicators for aviation and pesticide application rates or methane emissions from livestock for agriculture.

11 KPMG (2024a).

12 See GRI (2025) for more detailed background information, recent developments and comprehensive guidance on the application of the GRI Standards.

13 The majority of airlines publish sustainability reports based on the GRI Standards (see Johansson 2022 or SFC 2025). IATA (2024a) has aligned its Airline Sustainability Reporting Handbook (ASRH) with GRI to ensure consistency and comparability in airline reporting.

14 See SASB (2025) for more detailed background information, recent updates and access to the full set of industry-specific standards.

15 The SASB framework is also relevant for aviation due to its industry-specific approach and its focus on financial materiality, which aligns well with investor requirements in the sector. According to IATA (2024a), SASB provides sector-specific metrics for airlines and is explicitly considered in IATA's ESG metrics guidance to ensure comparability with investor-focused standards. Furthermore, SASB's Transport Sector Airlines standards include indicators such as the number of work stoppages or governmental enforcement actions related to aviation safety regulations, which are directly applicable to airline operations.

16 See TCFD (2025) for more detailed and latest information, including guidance documents, implementation examples and jurisdictional updates.

17 Several airlines have already adopted TCFD-aligned reporting to meet investor and regulatory expectations. According to IATA (2024a), TCFD helps investors assess how airlines mitigate climate-related risks, while the Smart Freight Centre (SFC 2025) notes that an increasing number of airlines integrate climate-risk disclosures and TCFD-based scenario analyses into their sustainability reports.

18 The IFRS (2025a) Foundation is an independent, non-profit organisation. It is best known for developing the International Financial Reporting Standards for financial reporting through the International Accounting Standards Board (IASB) that was expanded by the ISSB to create a global baseline for Sustainability Reporting.

19 See IFRS (2025b) for further details, recent updates and access to the full ISSB Standards.

20 While the SASB Standards remain in use, they are now managed by the ISSB and will gradually be integrated into future ISSB standards. The ISSB has also officially succeeded the TCFD in 2024, making it the key reference point for investor-focused Sustainability Reporting going forward.

21 According to IATA (2024a), the ISSB's emphasis on financially material and comparable indicators is particularly relevant for airlines aiming to address investor and regulatory requirements. Many aviation stakeholders are therefore preparing to align their reporting practices with ISSB Standards, which are expected to serve as the global baseline for investor-oriented ESG disclosures.

22 See CDP (2025) for detailed information on methodology, scoring updates and sector-specific questionnaires.

23 Several major airlines, including members of IATA, regularly disclose environmental performance through CDP (see for example IAG (2024), American Airlines

(2024) or Lufthansa (2024) for their latest CDP reports). According to IATA (2024a), CDP is considered a key investor-facing platform in aviation, particularly for demonstrating climate-risk management and progress in CO_2 reduction to financiers and regulators.

24 See GHG Protocol (2025) for detailed guidance, methodological updates and sector-specific calculation tools.

25 While the inclusion of indirect emissions represents a significant and meaningful extension of an industry participant's environmental footprint, it also increases the risk of double counting. For instance, the same aircraft emissions may be reported as Scope 1 emissions by the airline and simultaneously as Scope 3 emissions by the airport.

26 Many aviation stakeholders rely on the GHG Protocol for emissions accounting, as it provides the methodological basis for CO_2, NO_x and life cycle emissions calculations (IATA 2024a).

27 CDSB (2025).

28 UN (2025).

29 UNGC (2025).

30 PRI (2025).

31 ISO (2025).

32 See for example Carrots & Sticks (2020) and (2024) for evidence of the rapid growth and regulatory shift in Sustainability Reporting provisions worldwide, KPMG (2024a) for the increasing standardisation driven by investor and regulatory expectations and Reber, Gold and Gold (2022) for the strategic role of ESG disclosure in reducing information asymmetry and meeting investor demands.

33 KPMG (2024a) and GRI/SASB (2021).

34 According to KPMG (2024a), GRI is used by 71% of N100 and 77% of G250 companies. CDP is supported by over 740 investors representing more than 136 trillion USD in total assets. In 2023, over 23,000 companies disclosed data through CDP (CDP 2025). SASB is integrated into MSCI ESG ratings and 86% of S&P 500 companies report SASB-aligned (Saunders, McManus and Lockett 2023). TCFD is supported by more than 4,000 organisations in over 100 jurisdictions (TCFD 2025). 67% of S&P 500 companies publish TCFD-aligned disclosures (Saunders, McManus and Lockett 2023).

35 The ISSB Standards are being adopted or aligned with by numerous jurisdictions, including major financial markets such as Japan, Singapore, the United Kingdom and Brazil. Several securities regulators and stock exchanges are incorporating ISSB requirements into their listing and disclosure rules, establishing them as a foundation for national Sustainability Reporting frameworks (see IFRS 2024a; KPMG 2025).

36 See Carrots & Sticks (2024) and KPMG (2025).

37 For example, the UK's FCA requires TCFD-aligned disclosures for listed companies; Japan's FSA integrated TCFD into the Corporate Governance Code; New Zealand's Climate-Related Disclosures Act is fully TCFD-based, see TCFD (2022) and KPMG (2024a).

38 The original legislation is available online (EC 2025a). Additional resources include the European Commission's Explanatory Memorandum (EC 2021) and the Draft Commission Notice (EC 2024a). Further guidance and information on corporate Sustainability Reporting requirements in the EU are provided in EC (2025b).

39 The updated CSRD reporting requirements apply only to large undertakings with more than 1,000 employees and either a turnover above 50 million EUR or a balance sheet total exceeding 25 million EUR. This redefinition significantly narrows the scope of the directive, reducing the number of companies subject to Sustainability Reporting by approximately 80%.

40 KPMG (2024a).

41 The EFRAG is an independent technical advisory body of the European Commission, responsible for developing the ESRS. Its role is to ensure alignment with international frameworks such as ISSB and GRI while reflecting EU-specific policy goals (EFRAG 2025).

42 See KPMG (2023), KPMG (2024b) and IFRS (2024b) for more details.

43 EC (2024a).

44 In addition to the CSRD, aviation operators in Europe are subject to mandatory reporting under the EU emissions trading scheme, which was originally designed for emissions trading compliance. Starting in 2025, the EU will also introduce a dedicated MRV framework for non-CO_2 aviation effects. The reporting obligations of both initiatives will be discussed in more detail later in this chapter.

45 IATA (2024a).

46 The introduction of the CSRD further increases reporting requirements for aviation stakeholders operating in or with the EU. Given its extraterritorial scope, the CSRD will also affect non-EU airlines with significant operations or subsidiaries in the EU.

47 See, for example, Zieba and Johansson (2022), D'Silva (2025), Johansson (2022) and Hubbard (2025). These studies highlight substantial variation in reporting depth, structure and framework alignment, as well as persistent transparency gaps.

48 The following information is based on IATA's "Beginner's Guide to Airline Sustainability Reporting" (IATA 2024a). This guide serves as a simplified introduction to the development of an airline sustainability report and complements the more detailed IATA ASRH.

49 Including CSRD, ISSB, GRI, GHG Protocol, SASB and CDP.

50 Core metrics refer to well-established quantitative indicators that are already reported by most airlines or are relatively easy to obtain. Core+ metrics are less common, often reported by a smaller number of airlines, or more challenging to collect due to data availability or complexity.

51 IATA (2024a).

52 ACI (2023).

53 The ESG Management Best Practice document (ACI 2022) offers a global overview of widely used frameworks such as GRI, SASB and CDP, which are already commonly applied in the airport sector. Nevertheless, a sector-wide survey conducted by ACI revealed continued challenges, particularly the need for more tailored, airport-specific guidance and metrics.

54 See ACI-NA (2024). Although based on international frameworks and thus broadly applicable, the ACI-NA guidelines primarily address the North American context (ACI-NA 2024). Airports in Europe and North America share similar ESG priorities, but regulatory and ownership differences impacts their reporting practices. European airports operate in a stricter environment, with mandatory reporting under the CSRD and strong investor and societal pressure. In contrast, ESG reporting in North America remains largely voluntary and market driven, with only nine airports formally publishing ESG reports in 2023. Nevertheless, the ACI-NA guidelines can serve as a useful reference for airports globally, particularly those in earlier stages of developing ESG reporting structures.

55 The ESG Reporting and Metrics Task Group reviewed key sustainability and ESG disclosure guidance from the GRI, CDP, SASB, IFRS, TCFD, ISSB and ESRS. To ensure alignment with investor expectations, the Task Group also analysed ESG methodologies from major rating agencies (ACI 2024).

56 ACI-NA (2024).

57 ACA (2025).

58 Please refer to ICAO (2025a) for more details and latest information about CORSIA.

59 ICAO (2025).
60 See EC (2025c) for further details on the EU ETS and EC (2025d) for information regarding the MRV requirements for aircraft operators in this context.
61 To support aviation operators under the EU ETS, the EC (2025e) provides additional tools and guidance, including the free ETS Reporting Tool.
62 The following information is taken from EC (2024b), EC (2024c), EC (2025f) and EC (2025g). Readers are referred to these documents for additional insights into technical implementation and modelling approaches.
63 Operators are not required to provide Primary data but are encouraged to do so to enhance accuracy, as default values tend to overestimate climate impacts.
64 See, for example, IATA's letter to the European Commission (IATA 2024b) and Airbus's position paper "Considerations on the EU Non-CO_2 MRV Implementation" (Airbus 2025).

References

A4E – Airlines for Europe (2025) Assessment of the cost of regulatory compliance of European Airlines; https://a4e.eu/wp-content/uploads/Cost-of-doing-business-Final-Report.pdf

ACA – Airport Carbon Accreditation (2025) Airport Carbon Accreditation; https://www.airportcarbonaccreditation.org

ACI – Airports Council International (2022) Environmental, Social, and Governance (ESG) Management – Best Practice; https://store.aci.aero/product/environmental-social-and-governance-esg-management-best-practice

ACI – Airports Council International (2023) A global framework for improved airport ESG reporting; https://aci.aero/2023/10/31/a-global-framework-for-improved-airport-esg-reporting/

ACI – Airports Council International (2025) Policy Briefing; https://www.aci-europe.org/downloads/events/ACI%20EUROPE%20Policy%20Briefing%202025.pdf

ACI-NA – Airports Council International North America (2024) An Introduction to Performance and Risk Management Metrics for North American Airports; https://airportscouncil.org/wp-content/uploads/2024/07/ESGIntroductiontoPerformanceandRiskManagementMetricsforNorthAmericanAirports-051524.pdf

Airbus (2025) Considerations on the EU non-CO2 MRV implementation; https://www.lobbyregister.bundestag.de/media/62/d8/450239/Stellungnahme-Gutachten-SG2412180042.pdf

American Airlines (2024) CDP Corporate Questionnaire 2024; https://s202.q4cdn.com/986123435/files/doc_downloads/esg/esg-cdp-response-aag-2024.pdf

Carrots & Sticks (2020) Sustainability Reporting Policy: Global trends in disclosure as the ESG agenda goes mainstream; https://research.vu.nl/files/111479833/carrots_and_sticks_2020_interactive.pdf

Carrots & Sticks (2024) Recognizing the role of transparency to accelerate the SDGs – Annual Report 2024; https://www.carrotsandsticks.org/reports/2024-carrots-and-sticks-report.pdf

CDP – Carbon Disclosure Project (2025) Turning Transparency to Action; https://cdp.net/en

CDSB – Climate Disclosure Standards Board (2025) Climate Disclosure Standards Board; https://www.cdsb.net/what-we-do/reporting-frameworks

D'Silva, N. (2025) Towards Net-Zero: Analyzing Sustainability Reports of Global Airlines; in: *International Journal of Environmental Sciences*, Volume 11 (5s); pp. 781–813; https://theaspd.com/index.php/ijes/article/view/1268/984

Deloitte (2025) 2025 Gen Z and Millennial Survey; https://www.deloitte.com/content/dam/assets-shared/docs/campaigns/2025/2025-genz-millennial-survey.pdf

EC – European Commission (2021) Explanatory Memorandum; https://eur-lex.europa.eu/legal-content/EN/TXT/?uri=CELEX:52021PC0189

EC – European Commission (2024a) Draft Commission Notice; https://finance.ec.europa.eu/document/download/c4e40e92-8633-4bda-97cf-0af13e70bc3f_en?filename=240807-faqs-corporate-sustainability-reporting_en.pdf

EC – European Commission (2024b) Reference set of technical specifications for the MRV; https://climate.ec.europa.eu/document/download/735ae93d-d49d-46e0-b95b-48c36230ad57_en?filename=policy_ets_aviation_reference_set_technical_specifications_non_co2_mrv_en.pdf

EC – European Commission (2024c) A system for airlines to monitor, report and verify non-CO_2 effects of aviation – A step-by-step guide for airlines; https://climate.ec.europa.eu/document/download/2efafc7e-8b25-4763-906f-a7ba23b466d2_en?filename=policy_ets_avaiation_explainer_non-co2_mrv_tasks_for_ao_en.pdf

EC – European Commission (2025a) Directive (EU) 2022/2464 of the European Parliament and of the Council; https://eur-lex.europa.eu/legal-content/EN/TXT/?uri=CELEX:32022L2464

EC – European Commission (2025b) Corporate sustainability reporting; https://finance.ec.europa.eu/capital-markets-union-and-financial-markets/company-reporting-and-auditing/company-reporting/corporate-sustainability-reporting_en

EC – European Commission (2025c) About the EU ETS; https://climate.ec.europa.eu/eu-action/eu-emissions-trading-system-eu-ets/about-eu-ets_en

EC – European Commission (2025d) Aircraft operators and their administering countries; https://climate.ec.europa.eu/eu-action/eu-emissions-trading-system-eu-ets/monitoring-reporting-and-verification/aircraft-operators-and-their-administering-countries_en

EC – European Commission (2025e) ETS Reporting Tool (ERT); https://climate.ec.europa.eu/eu-action/eu-emissions-trading-system-eu-ets/monitoring-reporting-and-verification/ets-reporting-tool-ert_en

EC – European Commission (2025f) Reducing emissions from aviation; https://climate.ec.europa.eu/eu-action/transport-decarbonisation/reducing-emissions-aviation_en

EC – European Commission (2025g) MRV non-CO2 data collection Guidance for Aircraft Operators, Version 1.4; https://climate.ec.europa.eu/document/download/6aec6267-74ff-435a-87c7-92062db7ccb8_en?filename=policy_ets_avaiation_gd_non_co2_mrv_pre-neats_en.pdf

EC – European Commission (2025h) Commission simplifies rules on sustainability and EU investments, delivering over €6 billion in administrative relief; https://ec.europa.eu/commission/presscorner/detail/en/ip_25_614

Economidou, C., Gounopoulos, D., Konstantios, D., Tsiritakis, E. (2023) Is Sustainability Rating Material to the Market?; in: *Financial Management*, Volume 52 (1); pp. 127–179; doi:10.1111/fima.12406

EFRAG – European Financial Reporting Advisory Group (2025) Europe's Voice in Corporate Reporting; https://www.efrag.org/en

Fenili, A., Raimondo, C. (2021) ESG and the Pricing of IPOs: Does Sustainability Matter; https://ssrn.com/abstract=3860138

GHG Protocol (2025) GHG Protocol; https://ghgprotocol.org

GRI – Global Reporting Initiative (2025) Global Reporting Initiative; https://www.globalreporting.org/

GRI/SASB (2021) A Practical Guide to Sustainability Reporting Using GRI and SASB Standards; https://www.globalreporting.org/media/mlkjpn1i/gri-sasb-joint-publication-april-2021.pdf

Hubbard, S. (2025) Environmental Social Governance (ESG) Reporting for Large US Airports; in: *Sustainability*, Volume 17 (11); 4832; doi:10.3390/su17114832

IAG – International Airlines Group (2024) 2024 CDP Corporate Questionnaire 2024; https://www.iairgroup.com/media/bgxirlhp/iag-cdp-corporate-questionnaire-2024-responses.pdf

IATA – International Air Transport Association (2024a) Beginners Guide to Airline Sustainability Reporting Handbook; https://www.iata.org/contentassets/77ec9a8c8a864daaa00bdb7f5de02902/beginners-guide-to-airline-sustainability-reporting-april2024.pdf

IATA – International Air Transport Association (2024b) Proposed Monitoring, Reporting and Verification of Aviation Non-CO2 Emissions; https://www.iata.org/contentassets/21260d9fa3424714ab9cbf4c09bf55cf/letter-from-willie-walsh-to-evp-efovi-commissioner-hoekstra-and-commissioner-vlean.pdf

IATA – International Air Transport Association (2025) CORSIA Fact Sheet; https://www.iata.org/en/iata-repository/pressroom/fact-sheets/fact-sheet-corsia

ICAO – International Civil Aviation Organization (2025) Carbon Offsetting and Reduction Scheme for International Aviation (CORSIA); https://www.icao.int/carbon-offsetting-and-reduction-scheme-international-aviation-corsia

IFRS – International Financial Reporting Standards (2024a) Progress on Corporate Climate-related Disclosures – 2024 Report; https://www.ifrs.org/content/dam/ifrs/supporting-implementation/issb-standards/progress-climate-related-disclosures-2024.pdf

IFRS – International Financial Reporting Standards (2024b) ESRS–ISSB Standards – Interoperability Guidance; https://www.ifrs.org/content/dam/ifrs/supporting-implementation/issb-standards/esrs-issb-standards-interoperability-guidance.pdf

IFRS – International Financial Reporting Standards (2025a) Who we are; https://www.ifrs.org/about-us/who-we-are

IFRS – International Financial Reporting Standards (2025b) International Sustainability Standards Board; https://www.ifrs.org/groups/international-sustainability-standards-board

ISO – International Organization for Standardization (2025) ISO 26000; https://www.iso.org/iso-26000-social-responsibility.html

Johansson, E. (2022) An Analysis of Sustainability Reporting Practices of the Global Airline Industry; in: *Proceedings of the 5th International Conference on Tourism Research*, Volume 15 (1), pp. 507–516; doi:10.34190/ictr.15.1.234

Joseph, R. (2024) Investors and Regulators Demand Climate Disclosures, but Companies Should Beware of Greenwashing Claims; https://www.climatesolutionslaw.com/2024/09/investors-and-regulators-demand-climate-disclosures-but-companies-should-beware-of-greenwashing-claims

KPMG (2023) Comparing sustainability reporting requirements; https://kpmg.com/kpmg-us/content/dam/kpmg/frv/pdf/2022/comparing-sustainability-reporting-proposals-talkbook.pdf

KPMG (2024a) The move to mandatory reporting: Survey of Sustainability Reporting 2024; https://kpmg.com/xx/en/our-insights/esg/the-move-to-mandatory-reporting.html

KPMG (2024b) ESRS Foundations; https://kpmg.com/content/dam/kpmgsites/xx/pdf/ifrg/2024/isg-handbook-esrs-foundations.pdf.coredownload.pdf

KPMG (2025) Two years in: adoption of the ISSB Standards; https://assets.kpmg.com/content/dam/kpmg/ae/pdf-2025/06/two-years-in-adoption-of-the-issb-standards-en.pdf

Lufthansa (2024) 2024 CDP Corporate Questionnaire 2024; https://www.lufthansagroup.com/media/downloads/en/responsibility/LHG-CDP-Response-2024.pdf

PRI – Principles for Responsible Investment (2025) About the PRI; https://www.unpri.org/about-us/about-the-pri

PwC – PricewaterhouseCoopers (2024) Global Investor Survey 2024; https://www.pwc.com/gx/en/issues/c-suite-insights/global-investor-survey/global-investor-survey-report-2024.pdf

Reber, B., Gold, A., Gold, S. (2022) ESG Disclosure and Idiosyncratic Risk in Initial Public Offerings; in: *Journal of Business Ethics*, Volume 179; pp. 867–886; doi:10.1007/s10551-021-04847-8

Russell, B. (2024) NFRD vs CSRD: What's The Difference?; https://www.corefiling.com/2024/07/30/nfrd-vs-csrd

SASB – Sustainability Accounting Standards Board (2025) SASB Standards; https://sasb.ifrs.org/about

Saunders, E., McManus, L., Lockett, S. (2023) Framing It Up: Everything You Need to Know About ESG Framework Adoption Trends in 2022 and Beyond; https://riveron.com/posts/framing-it-up-everything-you-need-to-know-about-esg-framework-adoption-trends-in-2022-and-beyond

SFC – Smart Freight Centre (2025) 2024 SFC Airline ESG Survey – Benchmarking Report; https://smart-freight-centre-media.s3.amazonaws.com/documents/2024_SFC_Airline_ESG_Survey_Benchmarking_Report.pdf

Stepstone (2023) Three out of four employees would be more likely to apply to sustainable companies; https://www.thestepstonegroup.com/english/newsroom/press-releases/study-three-out-of-four-employees-would-be-more-likely-to-apply-to-sustainable-companies

TCFD – Task Force on Climate-related Financial Disclosures (2022) 2022 TCFD Status Report; https://www.fsb.org/2022/10/2022-tcfd-status-report-task-force-on-climate-related-financial-disclosures

TCFD – Task Force on Climate-related Financial Disclosures (2025) Task Force on Climate-related Financial Disclosures; https://www.fsb-tcfd.org

UN – United Nations (2025) Sustainable Development - The 17 Goals; https://sdgs.un.org/goals

UNGC – United Nations Global Compact (2025) Our Ambition; https://unglobalcompact.org/what-is-gc/mission

Zieba, M., Johansson, E. (2022) Sustainability reporting in the airline industry: Current literature and future research avenues; in: *Transportation Research Part D: Transport and Environment*, Volume 102; 103133; doi:10.1016/j.trd.2021.103133

8 Sustainability Strategies and Sustainability Management

Introduction

The aviation industry is undergoing a fundamental strategic transformation as sustainability becomes a decisive factor in corporate strategy development. Traditionally focused on cost efficiency, network optimisation and capacity growth, airlines are now required to integrate environmental considerations into their long-term strategic planning. Climate change, regulatory mandates and shifting stakeholder expectations increasingly shape the external business environment in which strategic decisions are made. In this context, strategy development is no longer limited to securing competitive positions in existing markets but also involves redefining value creation under environmental constraints. This chapter examines how sustainability – particularly through the adoption of Sustainable Aviation Fuel (SAF) – can be embedded into airline strategy, balancing risk mitigation with opportunity-driven approaches to ensure long-term competitiveness and legitimacy.

Fundamentals of Strategy Development in Aviation

The definition of mission and vision is the starting point for strategic corporate management. The mission describes the fundamental purpose and reason for the existence of a company. In some cases, it is also referred to as the company's purpose. It centres on the question "why do we do what we do?" and enables the company's drive and motivation to be defined. The mission describes the value proposition to stakeholders, the values that guide the company's actions and its role in the economy and the society. An example of this is DHL Group's mission: "Connecting people. Improving lives." DHL Group is one of the world's leading logistics providers, with approximately 600,000 employees in over 220 countries. It offers comprehensive solutions in the areas of express shipping, freight transport, supply chain management, e-commerce, mail and parcel services. The company's values are based on respect (e.g., health and safety of employees as a top priority) and results (e.g., mutual support and accountability). In contrast, the vision defines an ambitious picture of the future that serves as a reference

DOI: 10.4324/9781003608172-8

point for setting strategic objectives and prioritising measures. Both elements have a normative effect and provide the framework within which strategic decisions are made. Deriving a consistent corporate strategy from the mission and vision is not a linear process, but a reflective one that involves several stages.

Strategy translates the mission and vision into concrete medium- to long-term measures. It is the process by which companies define priorities to achieve the goals so that they can mobilise their resources. Strategy is essentially the creation of a unique and valuable position.[1] In a typically competitive environment, companies try to outperform their competitors by defining their own territory and creating value that customers recognise and are willing to pay for. Companies not only look for how to compete in markets with their rivals but also decide in which markets they want to operate and with which products. Firms develop their strategies based on a combination of internal and external analyses aimed at identifying competitive advantages, exploiting opportunities and minimising risks. Two aspects are decisive here: (A) the changes and trends taking place in the external business context and (B) the resources, skills and competences available within the company. Essentially, companies try to build on their own strengths and at the same time compensate for their shortcomings in order to mitigate threats that jeopardise their survival and to make the most of the opportunities that arise in the respective ecosystem as a business context.[2] For the strategy to be effective, it is essential that it is:

- Understood and lived internally (strategic alignment),
- Embedded in the corporate culture,
- Communicated coherently externally (e.g., via sustainability reports).

Shaping corporate strategy is not a one-off, static event. Rather, corporate strategy development is a dynamic process that requires constant adjustments as the external business context changes and the company's internal competences (e.g., innovations measured by patents) evolve. For many years, corporate strategy was primarily focused on economic success and detached from social and environmental issues. However, this has not been the case for many years, as social and environmental issues also dominate the public discourse. Companies are not only held responsible for causing social and environmental problems, but also for providing solutions.

Sustainability as a Strategic Driver in Aviation

The air transport industry is in the midst of a seismic change in its business context, both in the short as well as in the medium to long term. In the short term, the resilience of the system is evident in overcoming the aftermath of the COVID-19 pandemic. Air transport has recovered from the massive slump in demand between 2020 and 2022 on a global level.[3] However, trade barriers, economic slowdowns and geopolitical crises are creating new risks, the effects of which will mainly materialise in the medium term.[4] In addition, the world is facing the increasing consequences of climate change, which require prompt and consistent countermeasures. The aviation industry is confronted with several threats in connection with climate change:[5]

- Operational risks, as storms, heatwaves and extreme rainfall affect operations on the ground and, e.g., clear-air turbulence impacts flight operations,
- Reputational risks as customer preferences and perceptions of the environmental footprint of travel change,
- Financial risks, as investors and insurers lose interest in industries like the aviation industry that are dependent on fossil fuels,
- Regulatory risks, which are implemented through mandates and quotas by state and supra-national institutions (e.g., EU), often to the detriment of the industry,
- Future risks that might arise from voluntary commitments regarding emission reductions that have not yet been implemented to the extent announced.

In other words, there is a risk that fossil fuel-based industries might quickly lose their "license to operate".

Avoiding Greenhouse Gas (GHG) emissions takes centre stage in dealing with the consequences of climate change. Since companies do not operate in isolation, but within a larger supply chain network, they need to consider how they can avoid and reduce their GHG emissions both directly and indirectly. The GHG Protocol is a globally recognised standard framework for recording, balancing and reporting GHG emissions in companies, organisations and public institutions (see also Chapter 7). It was developed at the end of the 1990s by the World Resources Institute (WRI) and the World Business Council for Sustainable Development (WBCSD) with the aim of creating a standardised, transparent and comparable basis for emissions management. Today, the GHG Protocol serves as a central reference for climate accounting and forms the basis for numerous national and international regulations and Environment, Social and Governance (ESG) reporting standards. The GHG Protocol is a framework that supports companies to identify the different areas along a company's operations and supply chain network where GHG emissions could arise, known as Scope 1, 2 and 3 emissions (see also Chapter 7):[6]

- Scope 1: Emissions created directly by the company as a result of its own operations and manufacturing,
- Scope 2: Indirect emission from purchasing electricity e.g., for heating and cooling, for company's own use,
- Scope 3: Indirect emissions generated along supply chain activities both upstream (e.g., business travel) and downstream (e.g., processing of sold goods).

In the aviation industry, Scope 1 emissions represent the vast majority of GHG emissions. Every day, over 100,000 flights take off around the world. Global air traffic is forecast to double by 2043, fuelled by rising demand in Asia, Africa and the Middle East. Manufacturers such as Airbus and Boeing are forecasting demand for 44,000 new aircraft by 2043 (including replacements for old aircraft).[7] In view of the existing order backlog, which already extends into the 2030s, many of the low-emission aircraft ordered today will not fly until 2050 or later.

Against this background, it becomes clear that aviation is currently at a crucial point on its journey to sustainability and industry stakeholders will have little choice but to respond if they want to remain successful in the future. But wherever there are risks, there are also opportunities for companies. Those that adapt and reshape can pave the way to a business beyond the ordinary and profit in the process. Such opportunities also arise from tackling climate change by adopting a (circular) system or life cycle perspective along the aviation value chain. Industry players that proactively anticipate change and reorganise their business strategies can position themselves well for the future. However, such a transition must be carefully planned due to the long technological lead times in the aviation industry.

When it comes to sustainability and climate change, companies are required to integrate sustainability into business processes throughout their operations, including supply chains and partnerships. Sustainability issues affect almost every part of a company's operations: from the sourcing to production and manufacturing, from energy consumption and operational activities to emissions from those activities. In the context of the aviation sector, these could include:[8]

- Raw material sourcing: extraction of metals for airplane manufacturing; agricultural practices of food production sourced for onboard catering,
- Production and manufacturing of investment and intermediate goods: energy use in manufacturing airplanes; food waste in putting together meals,
- Energy use of infrastructure providers: Airports and maintenance facilities consume large amounts of electricity, which often does not yet come from renewable sources,
- Logistics: transport of parts and components based on global supply chains; relocation of crew to serve on flights; flight routes,
- Operational activities of airlines: fuel emissions of flights; food and plastic waste from onboard services,
- Consumer behaviour: emissions caused by passengers travelling to and from airports.

In order to integrate sustainability into their business processes, companies in the aviation sector have moved away from linear economic models (take-make-waste) in which they take what they need from the natural environment and produce emissions without considering the negative external effects. Instead, processes are increasingly aligned with a circular model, where the use of both biological and technical materials is designed to avoid waste and pollution. Technical (man-made) materials are kept in the system through measures such as sharing, maintaining and extending, reusing and redistributing, refurbishing and remanufacturing or recycling products. The aim is to minimise waste and to reduce the use of non-renewable resources that causes negative external effects. By adopting a circular approach, companies can create marketable values through pollution prevention, product stewardship, clean technology and collaboration throughout the supply chain, including local communities.[9] Competitive advantages can also be achieved by taking a pioneering role and building innovation expertise through the integration of sustainability solutions. This also includes anticipating regulatory requirements.[10]

Stakeholders in Aviation and Management of Emissions

There are four main sectors within the aviation industry: airports and ground handling companies (GHCs), air navigation service providers (ANSPs) and airlines.

Airports are among the most energy-intensive infrastructures in the aviation sector. Terminal operations, energy supply and passenger mobility all cause ecological impacts. Ground handling processes account for a significant proportion. Activities like the use of ground service equipment, apron logistics, baggage handling and aircraft services are directly linked to energy consumption and emissions. Against this backdrop, GHCs are an integral part of airports' holistic sustainability strategies. Airports therefore rely on comprehensive sustainability strategies to reduce their ecological footprint. One key area of action is the reduction of GHG emissions through the use of renewable energies and energy-efficient technologies. When expanding airports, increasing attention is being paid to environmentally friendly materials and resource-saving construction methods. As an example, Oslo Airport has used only recyclable steel, volcanic ash-based concrete and certified wood in its 2018 terminal extension. In addition, snow collected in winter is used to cool the building in summer.[11] Airports also focus on circular economy and waste avoidance or separation. For example, Changi Airport (Singapore) operates a state-of-the-art waste separation system and uses rainwater to irrigate its green spaces.[12]

Air Traffic Management (ATM) can contribute meaningfully to sustainable aviation, not primarily by structural reorganisation, but also through operational excellence and digital intelligence. While projects like SESAR (Single European Sky ATM Research) have resulted in innovations such as:

- Free Route Airspace: Allows airlines to use direct flight routes between freely selectable entry and exit points,
- Continuous Descent Operations and Continuous Climb Operations: Optimised climb and descent procedures that are more energy efficient while reducing noise emissions near airports (see also Chapter 6 for a detailed description).

Environmental gains, e.g., in Europe, depend less on the defragmentation of airspace and more on real-time decision support, trajectory-based operations and seamless air-ground data exchange.

A promising future lies in the latency-free, secure communication of intent and performance data. This will enable multi-criteria flight path optimisation in which safety, efficiency, noise, contrail formation and fuel consumption are taken into account in a balanced manner. This shift from "optimising the average" to "optimising the individual flight" represents the next stage in the development of environmental responsibility of air traffic management.[13]

While all stakeholders add to the overall ecological balance, airlines play a key role. They are the most important lever for sustainability both in quantitative and systemic terms. Airlines are responsible for the majority of GHG emissions in aviation. Around 80–90% of CO_2 emissions in aviation are generated during the flight and therefore directly by the airlines' flight operations.[14] Airports, ATM and ground handling also cause emissions, but to a much lesser extent.

Airlines are at the centre of the aviation-related value chain and can act as demand drivers for sustainable innovations. They influence both aircraft manufacturers through the demand for more efficient aircraft as well as airports and ANSPs through cooperation requirements with regard to sustainable operating procedures. One example of cooperation between airline, airport and ANSP is the "Collaborative Environmental Management" (CEM) project at London Heathrow Airport. British Airways, the airport operator Heathrow Airport and NATS as UK ANSP are involved among others. Measures include optimised approach and departure procedures (e.g., Continuous Descent Operations), the identification of inefficient flight profiles, coordinated airway slot and ground handling planning to minimise waiting times with engines running and sustainability KPIs, which are evaluated jointly.[15]

International Airline Climate Targets

In 2010, the International Air Transport Association (IATA), whose 340 member airlines nowadays handle 80% of global air traffic, set three targets for reducing CO_2 emissions:[16]

- Fuel efficiency, by an average improvement of 1.5% per year (2009 to 2020),[17]
- Carbon-neutral growth, by a cap on net CO_2 emissions from 2020 onwards,
- Reduction in net CO_2 emissions, by 50% by 2050, based on 2005 levels.

In 2021, a resolution was passed to achieve net zero carbon emissions by 2050 to support the goal of the Paris Agreement. In order to achieve these targets, the industry is pursuing a so-called "basket of mitigation measures", which consists of four pillars and involves all stakeholders of the aviation sector (see also Chapter 6):[18]

- Technology, e.g., development of more fuel-efficient aircraft and advancement of SAF, such as Methanol-to-Jet,
- Operations, e.g., enhanced ground operations (e.g., reduced taxi time),
- Infrastructure, e.g., optimisation and shortening of flight routes and procedures,
- Market-Based Measures, e.g., Carbon Offsetting and Reduction Scheme for International Aviation (CORSIA – see also Chapter 5), carbon pricing schemes, voluntary carbon offset programmes.

The Long-Term Aspirational Goal (LTAG) of the International Civil Aviation Organization (ICAO) was adopted at the 41st General Assembly in 2022 (see also Chapter 6). The LTAG aims to reduce global net CO_2 emissions from international aviation to zero by 2050 taking into account the different economic conditions and technological paths of the member states. This is the first time that a globally coordinated, long-term target framework for the decarbonisation of the sector has been established, based on a broad multilateral consensus. The ICAO is thus positioning itself as a climate policy steering body that influences the strategic direction of aviation through legally non-binding but politically relevant targets. Compared to market-driven initiatives such as the net zero targets of major airlines, the LTAG offers a reference framework based on international law that also structurally integrates emerging and developing countries. It also takes into account the different levels of development of the member states and emphasises the role of technological cooperation, financing mechanisms and capacity building. The LTAG expands the regulatory horizon beyond market-based measures such as CORSIA and calls for the systematic integration of technological innovation, international cooperation and cross-sectoral climate strategy. It provides a normative framework for implementation at national and regional level, e.g., in the form of harmonised SAF quotas, infrastructure funding or research partnerships. The LTAG is not yet linked to mandatory reduction pathways or a global emissions trading system but is based on voluntary contributions by countries similar to the principle of "Nationally Determined Contributions" (NDCs) under the Paris Agreement.

Technological progress, especially in the engine sector, and the use of SAF are among the most important cornerstones to achieve net zero. The initially high importance of CO_2 compensation programmes is increasingly being called into question as compensation schemes have multiple limitations.[19] Only a small percentage of passengers offset their emissions. In addition, many other industries want to offset their emissions, but there are not enough projects to cover all sectors. Offsetting is often carried out in developing countries, which not only raises ethical issues, but also questions of monitoring and control of the projects. Initiatives such as reforestation can often take decades before CO_2 emissions are fully neutralised. And so airlines react accordingly. Ryanair discontinued its own CO_2 compensation programme in May 2025, and Wizz Air no longer actively offers its own.

The purchase and deployment of modern, fuel-efficient aircraft is in line with airlines' fundamental endeavours to reduce operating costs and increase profitability by using lower-emission aircraft. However, the long-term delays in delivery by aircraft manufacturers Boeing and Airbus are creating increasing barriers to achieving sustainability targets. Other propulsion technologies such as electrically powered or hydrogen-fuelled aircraft will only come into use in the 2050s and will only then have an effect on the intended reduction in emissions (see Chapter 6).

A lack of major progress in the area of operational efficiency and emerging regulatory requirements in Europe (Fit For 55, ReFuelEU Aviation, increasingly also in other parts of the world) have led to the use of SAF becoming a key element of decarbonisation efforts.

Sustainable Aviation Fuel as a Strategic Lever

Conventional Aviation Fuel and Emissions

Today's jet fuel is the result of military developments during World War II. With the invention of the first generation of jet engines for fighter aircraft, the first jet fuel was certified as JP-1 (Jet Propellant 1). The thermal and storage stability of jet fuel enabled the growth of civil aviation as the risk of quality deterioration during transport or storage was limited. With the introduction of jet engines in passenger aircraft in the early 1960s, airlines were able to reach higher altitudes and fly at higher speeds. The increase in traffic in the 1960s eventually led to specifications with multiple requirements:

- Must comply with international standards (ASTM D1655, DEF STAN 91–091),
- Must be affordable, available worldwide, easy to store,
- Usable without modification of aircraft or airport infrastructure,
- High energy density with the lowest possible mass,
- Freezing point of -47° Celsius in order to be able to fly up to an altitude of 40,000 feet,
- High flash point and good vaporisation in the engine's combustion chamber.

This specification is summarised under the type designation JET A-1. However, only minimum or maximum values or ranges are defined for the fuel characteristics. As a result, there is no standardised production technique for obtaining aviation fuel. Due to different processing technologies and different crude oil sources, the chemical composition of each batch of JET A-1 varies slightly.

Aircraft manufacturers specify which type of jet fuel may be used in their aircraft. Regulatory authorities, in turn, require airlines to use only the specified fuel. However, as the fuel is not an integral part of the aircraft, the production, distribution and storage of JET A-1 in the supply chain upstream of the aircraft is not covered by regulation.

There are no government regulations for fuel quality and quality monitoring. Major oil companies have started to endorse voluntary standards for the production and handling of aviation fuel. In addition, there are regular inspections of refineries, fuel depots and refuelling services at airports, to monitor compliance with these standards. Following EU Directive 965/2012, the IATA airlines established a voluntary technical review group (IFQP, the IATA Fuel Quality Pool).

At large airports, aircraft are refuelled via an underground hydrant system that is connected to the airport's fuel storage facility via pipelines. This storage facility is operated under mixed storage conditions. All supplies of jet fuel from different sources that comply with the JET A-1 specification can be stored together, as the same blending also applies to aircraft fuel tanks containing residual fuel from previous refuelling operations. Aircraft at small and medium-sized airports are supplied by refuelling vehicles, which are loaded at the airport fuel depot and pump the fuel from their vehicle-mounted tank to the aircraft.

Airlines report various key figures on their transport capacity and load factors. However, actual data on fuel consumption is only available on an aggregate level. There is currently no "single source of truth" on global aviation fuel consumption. As described in Chapter 4, around 942 million tonnes of CO_2 were emitted by global air traffic in 2024, and 1 kg of fuel leads to about 3.15 kg of CO_2 emissions. Therefore, around 300 million kg of fuel were burned. This figure marks an exceed of pre-crisis levels from 2019 after consumption plummeted in 2020 due to the COVID-19 pandemic. Demand for jet fuel is expected to more than double by 2050. Fuel consumption is rising sharply due to increasing air traffic. At the same time, the growth in emissions is being slowed by efficiency improvements and the increasing use of SAF. The existing range of uncertainty results primarily from different assumptions regarding fuel efficiency improvements and the share of SAF.

The price of JET A-1 is not only determined by production costs, but also by market conditions. In addition to common determinants such as quality, quantity and availability, the "scarcity rent", i.e., the premium that consumers are willing to pay on the spot market for immediate availability, also comes into play.

As fossil resources become increasingly scarce, oil production is shifting to more cost-intensive regions (e.g., oil sands). Production costs for such sources peak at up to USD 70 per barrel, leading to a rising price level for JET A-1 in the long term.[20]

The market for jet fuel represents a crucial supply structure within the global aviation industry. Due to high safety and quality requirements as well as limited substitutability, the jet fuel market is largely oligopolistic with vertically integrated supply chains. A few key players exercise extensive control over pricing, infrastructure and supply conditions. Global production is primarily carried out by large oil companies, including Shell PLC, Exxon Mobil Corp and BP PLC (Air BP). These firms generally cover all stages of the value chain from crude oil production (upstream) and refining (midstream) to distribution and airport supply (downstream). Another part of jet fuel is traded via international commodity traders such as Vitol, Glencore or Gunvor. These companies utilise arbitrage options between regions. Global trade takes place primarily via sea transport, using the hubs of Singapore, Rotterdam and Houston in particular. The jet fuel market therefore has the characteristics of an international commodity market with strong price links to spot and futures markets. As the final stage of the supply chain, direct fuelling at the airport (into-plane fuelling) is carried out either by specialised service providers such as Skytanking or Menzies Aviation or by subsidiaries of oil companies themselves. In many cases, airports operate on the basis of a so-called fuelling consortium, in which airlines also participate in the financing and utilisation of the infrastructure (storage, pipeline, hydrant systems). In some cases, an airline can become a fuel supplier and is therefore subject to the obligations arising from the European ReFuelEU Aviation requirements. According to EU law, a distributor is a natural or legal person who distributes fuel within the territory of a member state, i.e. releases it for use (Article 2 REDII/III). Airlines are fuel suppliers if a fuelling service provider is directly contracted and commissioned and the delivery is not secured "into-plane" by an oil company.

In the USA, stable national transport networks enable long-term volume contracts to be awarded, which tend to lead to more favourable conditions for offtakers. Airlines are also increasingly using financial products to hedge price risks (e.g., caps). These instruments make it possible to limit cost volatility, but do not replace the necessity of well-founded market analyses and a well-thought procurement strategy.

Definition and Characteristics of Sustainable Aviation Fuel

SAF is regarded as a key technological lever for decarbonising the aviation industry. SAF is a liquid fuel currently used in commercial aviation that reduces CO_2 emissions by up to 80–90%. Designed as a "drop-in" fuel, SAF can be used in existing turbine aircraft, but offers significant advantages in production, regulation and environmental impact compared to fossil fuel.

The first successful demonstration flights with SAF took place in 2008. Since then, SAF has developed from an experimental product to an industrially produced and increasingly regulated commodity. By the end of 2024, around 700,000 flights were operated with SAF, mainly with a blend of up to 50%. Although SAF is designed as a drop-in fuel and is therefore technically compatible with fossil JET A-1 fuel during flight operations, it is a commodity in its own right from a market, production and regulatory perspective.

SAF is produced from sustainable, non-fossil raw materials or using synthetic processes currently only in small quantities (see also Chapter 6). Production therefore differs fundamentally from the fossil refining of jet fuel. It can be produced from a range of sources (feedstocks), including waste oils and fats, green and municipal waste and non-food crops. It can also be made synthetically through a process that captures carbon directly from the air. It is "sustainable" as the feedstock does not compete with food crops or water supplies or contribute to forest degradation. While fossil fuels increase CO_2 emissions by releasing previously sequestered carbon, SAF "recycles" the CO_2 absorbed by the biomass of the feedstock during its lifetime.

Nine process technologies are certified for the production of SAF, which are equivalent in performance to JET A-1 fuel. The most important pathways are:

- HEFA-SPK (Hydroprocessed Esters and Fatty Acids – Synthetic Paraffinic Kerosene): Utilisation of used cooking oils and animal fats,
- FT-SPK (Fischer-Tropsch Synthetic Paraffinic Kerosene): Biomass gasification with Fischer-Tropsch synthesis,
- AtJ-SPK (Alcohol-to-Jet Synthetic Paraffinic Kerosene): Conversion of alcohol into jet fuel,
- PtL-SPK (Power-to-Liquid Synthetic Paraffinic Kerosene): Synthesising jet fuel from green hydrogen and CO_2.

SAF products are subdivided into two main categories:

- Fuels that do not fully meet the specifications of JET A-1, e.g., due to a lack of aromatics or insufficient density. These are only authorised as synthetic blending components (SAF-SBC),
- Fuels that meet the full JET A-1 specification. For regulatory reasons, these can currently also only be used as a 50% blend.[21]

SAF Certification Requirements and Regulatory Framework

SAF enables a significant reduction in GHG emissions over the entire life cycle (up to 80–90% compared to fossil jet fuel). This property gives SAF a unique environmental and economic value.

However, the sustainability of these processes depends heavily on the origin of the feedstocks used. Waste and residual materials such as used oils or forestry residues are considered to be particularly climate friendly, while energy crops can potentially cause Indirect Land Use Change (ILUC).

SAF is subject to specific regulatory and voluntary sustainability requirements and certification obligations (e.g., CORSIA). These regulations do not apply to fossil jet fuel and affect both the market presence and eligibility of SAF. Airlines must integrate the procured SAF quantities into their carbon footprint accounting and ensure traceability.

ICAO's CORSIA, which is not yet mandatory, defines criteria for the crediting of SAFs to reduce emissions (see also Chapter 5). Certified SAF must achieve GHG savings of at least 10% compared to fossil jet fuel.

In recent years, a lack of progress in the decarbonisation of aviation has led to stricter regulation aimed at promoting the use of environmentally friendly fuels in the aviation industry. While the EU is focusing on binding targets (mandates) from 2025, the USA is pursuing a different approach. This is intended to ensure a greater admixture of SAF through corresponding subsidies (see also Chapter 5).

The EU has introduced binding blending quotas as part of the "ReFuelEU Aviation" package: A minimum of 2% SAF is required from 2025, with an increase to 70% by 2050, introducing a separate sub-quota for electricity-based PtL SAF (see Chapter 5).

The USA promotes SAF via the Inflation Reduction Act (IRA) with tax credits, among other things. In addition, funding programmes such as the SAF Grand Challenge are driving the scaling of corresponding systems (see also Chapter 4).

SAF has made great progress since 2008 from first test flights to commercial plants, supported by certification schemes (ISCC, CORSIA, EU Renewable Energy Directive (RED)), regulatory quotas (ReFuelEU) and growing global production. Nevertheless, market penetration remains low and key challenges such as production costs, feedstock availability, scalability risks of different production pathways and the need for infrastructure adaptation have to be addressed.

Market Structure, Availability and Scaling of SAF

While fossil jet fuel is traded globally as a standardised, highly liquid bulk commodity, trading in SAF has so far mainly taken place within a framework of specific purchase programmes. The price is significantly higher, which is due to limited availability, complex production processes, lack of scaling and additional certification requirements. The market for SAF is currently only liquid to a limited extent and financial instruments for hedging risks, so-called SAF hedging, have hardly been developed.

Global SAF production capacity in 2025 has reached around 1.9 million tonnes, which corresponds to less than 0.6% of global jet fuel consumption.[22] HEFA production currently dominates with a share of 99%, as PtL plants are largely still in the planning phase. No large-scale eSAF plant with a production capacity of over 10,000 tonnes per year is yet in operation. There are 48 eSAF projects under development in Europe (as of December 2025). The envisaged production capacity is equivalent to three times the annual volume required by ReFuelEU by 2032. None of these projects has reached a Final Investment Decision (FID). While 11 large-scale plants are under development in China, the first FID worldwide was secured for a large-scale project in the USA (Infinium, Texas). By 2030, SAF capacity could increase to over 10 million tonnes if all announced projects come on stream. The share of SAF produced from HEFA will decrease to 80% in 2030, as a number of alternative processes will be used. Methanol-to-Jet can also be considered as a way to produce eSAF if e-methanol is used as a feedstock. However, as of the end of 2025, this route has not yet been approved by the American Society for Testing Materials (ASTM) for use in commercial aircraft. SAF is currently supplied by a small number of specialised producers (e.g., Neste, World Energy, Gevo) and is currently only available at selected locations. Despite costly certification in accordance with international SAF standards (e.g., ASTM D7566), SAF producers are increasingly turning to utilising the existing refinery infrastructure in order to enter SAF production more cost-effectively in the so-called co-processing technology. SAF producers generally have the option of switching to the production of renewable diesel and utilising similarities in raw materials, infrastructure and processing technologies. The HEFA route enables the co-production of SAF and Hydrotreated Vegetable Oil (HVO) diesel using biogenic raw materials such as used cooking oil or animal fats. From a technical perspective, modular refinery plants allow producers to customise production fractions through distillation and blending. The rationale for such a shift includes broader demand profiles, stable regulatory frameworks for road fuels and comparatively mature market incentives such as blending mandates and tax credits. Renewable diesel markets offer volume scalability, lower volatility and established distribution channels, potentially enhancing capital efficiency for producers navigating SAF cost pressures and policy uncertainty. Regulatory differentiation under directives like EU RED II requires separate sustainability certifications and life cycle assessments and could therefore pose compliance challenges.

The current demand for SAF is determined by the interplay of regulatory obligations and voluntary commitments. Regulatory commitments are typically structured as a mandate requiring either (or both) a reduction in the carbon intensity of the total fuel used or the use of a certain amount of low-carbon fuels. The voluntary demand for SAF by airlines, which exceeds legal requirements, is a result of strategic, market-oriented and climate-related considerations. Numerous airline (groups), such as Air France–KLM, have committed to net zero targets by 2050. SAF is a bridging technology until alternative propulsion systems (e.g. hydrogen) become available and is used as a key instrument for reducing airlines' Scope 1 emissions. A significant proportion of the voluntary demand for SAF is the result of pressure from corporate customers who are themselves pursuing ambitious climate targets. Multinational corporations such as Microsoft have their own net zero strategies and are subject to reporting obligations (e.g., GHG Protocol). Airlines are therefore developing specific offers such as cooperation models (e.g., Lufthansa "Corporate SAF Partnership") to meet customer-driven demand. However, the voluntary use of SAF can also serve as a differentiating factor in the competition for customers and investors. In a climate-orientated environment, the trustworthy procurement and use of SAF strengthens the company's image. The timely implementation of voluntary SAF procurement schemes also serves to secure long-term supply assurance and price conditions. In view of limited SAF production and increasing quota requirements, access to reliable supply volumes is a strategic advantage.

Global demand for SAF continues to grow, although many SAF projects in operation have to cope with the harsh reality of early market development. While the political signals are generally strengthening and the announcement of purchase commitments by companies continues, in 2024 and also in 2025 more production capacity became inactive than newly announced in some regions. Fewer than expected long-term commitments from the voluntary market have made some projects difficult to justify. Biogenic SAF produced from natural oils or biomass has become the preferred technology for many airlines due to its relative maturity and faster time to market. Despite its promising properties, its introduction faces another major hurdle, with SAF prices significantly higher than conventional jet fuel airlines face the challenge of reducing their environmental impact while maintaining their economic viability.

Pricing, Contract Models and Economic Risks of SAF

Despite growing political support and technological progress, price formation for SAF has not yet been standardised or consolidated within a transparent market structure. SAF is a new commodity and due to the low volumes of production and the lack of a stable market structure there is no spot pricing available. Instead, sales are handled on a bilateral basis. This means that different contract structures are used, each with their own strengths and risks. There are three main types of contracts:

- Indexed to commodity price: the most common contract structure is where SAF is sold at a fixed premium to an established commodity, usually jet fuel or crude oil. This type of contract is attractive to buyers as it allows the price of SAF to follow conventional jet fuel prices, which airlines are aware of and know how to hedge against. For producers, however, this can be a challenge as their revenues are linked to the volatility of the oil price, while their costs are determined by commodity prices.
- Cost-plus: A fixed mark-up on the SAF production costs is agreed. This type of contract does not depend on market dynamics, such as the demand for SAF or jet fuel. Rather, the price depends on the production costs. These include feedstock costs (including purchasing and logistics), process costs (refining, energy consumption, storage), downstream costs (transport to blending or airport storage, fuelling service), administration, marketing and risk mark-ups. The price components vary considerably depending on the type of contract (e.g., ExWork, Free on Board, Cost-Insurance-Freight, free into plane). This contract secures revenue for

producers, which is particularly attractive for those who want to develop projects and attract investment. The buyers bear a higher risk as they are now exposed to the volatility of input factors such as electricity, hydrogen and biomass, which is in most cases a new market for them.

- Fixed price: Some contracts are set at an agreed value that does not change over time. They offer security to buyers and sellers but can also put the buyer in an unfavourable position if the market price of SAF or jet fuel falls. Typically, fixed price contracts are used when the sale is for a limited quantity over a short period of time.

Demand Development: Regulated vs. Voluntary SAF Demand

SAF prices vary depending on the region to which the SAF was delivered. In the USA and also in Europe, SAF prices tend to be lower, while prices in the Asia-Pacific region are significantly higher on average. The difference is due to the political support in each region.

While North America and Europe have comprehensive policy frameworks, they pursue different approaches. In the USA, SAF production is stimulated through financial incentives, but there is no policy mandating the use of SAF. In 2021, the government announced the SAF Grand Challenge, which aims to achieve SAF consumption of three billion gallons by 2030, but is not legally binding.

In Europe, on the other hand, the focus is on mandates. Both the UK SAF and ReFuelEU Aviation mandates started with 2% of jet fuel demand in each region in 2025. The SAF price in Asia-Pacific is higher than in North America and Europe. Some countries have now announced SAF policies to be implemented by the end of the decade. The most established regional policy is in Singapore. The Singaporean government has announced a 1% SAF blending on outbound flights from 2026.

To summarise the market penetration of SAF depends crucially on five factors:

- The global development of the price for oil and conventional jet fuel,
- The price of allowances within an ETS,
- The introduction of binding mandates and quotas (e.g., ReFuelEU Aviation),
- Governmental subsidies,
- Access to private investment capital.

The sustainability debate in aviation has so far been strongly coloured by a Western industrial perspective. The narrative is dominated by large network airlines, European and North American regulation and technology-advanced manufacturers. However, this focus largely ignores the role of smaller players such as regional airlines or airports other of global hubs as well as the conditions and challenges in emerging aviation markets. A realistic and fair transformation of air transport requires the inclusion of these stakeholder groups, as they play a central role in growth, connectivity and social participation in many regions of the world. While large network airlines are increasingly entering into SAF offtake agreements or corporate SAF partnerships, regional airlines often lack access to financing instruments or political support that would enable a sustainable transformation.

In developing countries and emerging markets (e.g., India, Brazil), air transport is a crucial component of the national infrastructure, as alternatives such as high-speed trains are lacking. Air transport not only fulfils economic functions there, but also a social integration task, for example in medical care or in connecting remote regions. The introduction of highly regulated sustainability standards such as mandatory SAF quotas or complex emissions trading schemes would lead to massive distortion and marginalisation if the corresponding technological or institutional prerequisites are missing. There is also a global imbalance in the distribution of SAF production capacities, investments and certification structures. While political framework conditions such as ReFuelEU or the Inflation Reduction Act (IRA) create targeted incentives in Europe and North America, many countries lack the political capacities, regulatory institutions or investors. This asymmetry can lead to "two-speed aviation", with sustainable technologies and processes being introduced primarily in highly developed markets, while many emerging and developing countries fall behind. While Europe and North America are accelerating the use of SAF and more efficient fleets through regulation and subsidies, other regions often lack the financial resources and infrastructure to do so. The result is not only increasing distortion of competition in global air transport. There is also the threat of negative effects on the global emissions balance, if traffic flows are shifted to regions with lower environmental standards (carbon leakage). From an emissions and fairness perspective, this development is particularly problematic. Countries that have historically contributed little to cumulative emissions run the risk of bearing a disproportionate share of the adaptation costs without benefiting to the same extent from technology transfer or new value chains in sustainable aviation. Without coordinated international compensation mechanisms, the decarbonisation of aviation could therefore exacerbate existing global inequalities rather than reduce them. An inclusive sustainability policy in aviation must therefore consider differentiated paths and transition periods that recognise regional realities. This includes technology transfer, international climate financing, adaptation of certification standards, but also the funding of regional SAF production, taking into account local resources and socio-economic conditions. Global institutions such as ICAO or multilateral development banks (e.g., World Bank, African Development Bank) could take on a coordinating role here.

The sustainable transformation of aviation is only realistic and legitimate if it not only takes into account the Global North, but also actively integrates smaller players and high-growth regions of the Global South.[23]

Risk Mitigation versus Business Opportunity as Strategy Poles

Airlines play a key role when it comes to environmental sustainability in the aviation industry. Not only do they cause massive Scope 1 GHG emissions as part of their day-to-day operation, they also face increasing pressure from stakeholders on climate change action. There are two fundamental strategic approaches for airlines. Both "Business Opportunity" and "Risk Mitigation" are strategic options that airlines can exploit in connection with sustainability. Both strategies take into account regulatory requirements and social expectations and have the potential to reduce the negative environmental impact caused by air transport (see Figure 8.1).

These strategies have separate purposes but are complementary as many companies pursue a combination of both strategies. However, they differ in their focus and objectives as well as time horizon. While "Business Opportunity" aims to use sustainability as a competitive advantage, "Risk Mitigation" focuses on minimising the negative effects of sustainability risks.

From a risk perspective, managing climate change is about avoiding the loss of a company's reputation. A risk perspective usually focuses on the demands of interest groups such as investors, political decision-makers, customers, the media and activists. As part of a risk mitigation strategy, companies lower their unsystematic risk and reduce the costs associated with emissions and violations of regulatory requirements. On the other hand, it also prevents a loss of reputation and the resulting potential loss of legitimacy.

From an opportunity perspective, addressing climate change creates value for the organisation by building internal capabilities (e.g., procurement, technology, production) that create a competitive advantage by reducing costs, staying ahead of competitors and creating a leading position in the industry.[24] Companies that adopt this perspective have stronger brands, a more favourable cost position through operational efficiency, higher customer loyalty, additional revenue streams and higher employee productivity (see Table 8.1).[25]

In the face of the growing pressure from stakeholders like customers and competitors, it is clear that a long-term strategy, which includes sustainability, is

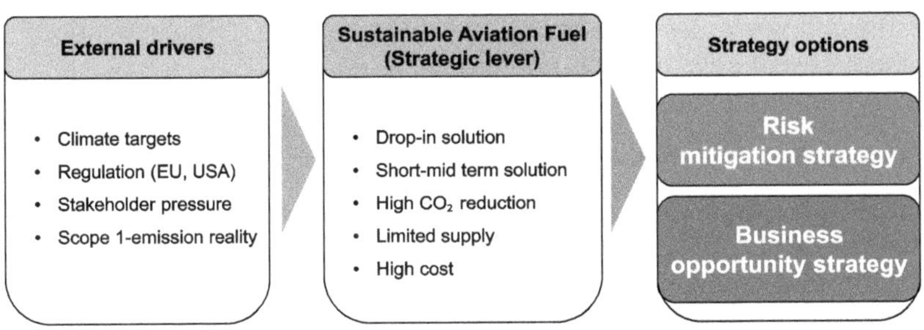

Figure 8.1. Strategy options risk mitigation and business opportunity.

Table 8.1. Aspects of risk mitigation and business opportunity strategy.

Aspect	*Risk mitigation*	*Business opportunity*
Objective	Minimise risks through sustainability requirements, ensure financial stability, maintain operating license	Exploit opportunities through sustainable measures, creating competitive advantages
Focus	Regulations, cost control, reputation protection	Innovation, new markets, customer loyalty
Financial investment	Low	Medium – high
Measures	Adaptation to CO_2 regulations, reduction of ecological footprint to avoid penalties	Long-term contracting of sustainable fuels, new business instrument such as Green Fares, SAF surcharges
Time frame	Short- to medium-term, often reactive	Long-term, proactive

an essential, indispensable element for long-term success. The need for urgent action is reinforced by a paradox in aviation. While CO_2 emissions per passenger kilometre are on the decline, the industry's absolute CO_2 emissions continue to rise (see Chapter 2).

Airlines often use a mixture of both strategies in order to secure competitive advantages on the one hand and avoid financial and legal risks on the other. SAF is crucial for this, as it is widely regarded as the most effective option for decarbonising aviation. The aviation sector will need around 500 million tonnes of SAF annually by 2050 to achieve the net zero CO_2 emissions target. Due to the availability of sustainable raw materials and technological maturity, global SAF production is expected to reach only around 400 million tonnes by 2050. The structural supply gap of around 100 million tonnes is due, among other things, to the pace of technology introduction beyond currently mature processes such as HEFA.[26] New processes, including Fischer-Tropsch and PtL, are not yet sufficiently scaled. From a strategic management perspective, this quantified supply gap transforms SAF from a purely ecological solution into a scarce strategic resource, reinforcing the need for strategy development. Using the example of the procurement of SAF, the following section shows how the strategies "risk mitigation" and "business opportunity" can be implemented.

Implementation of a Risk Mitigation Strategy

In addition to environmental benefits, the introduction of SAF also entails operational and economic risks. In order to minimise these risks, many airlines rely on a targeted risk mitigation strategy without taking advantage of the opportunities arising from a "green transformation" for the company.

Adaptation to CO_2 regulations is primarily undertaken in the short to medium term as part of a risk mitigation strategy and is often reactive. The target is to reduce the ecological footprint and avoid financial penalties due to non-compliance with legal requirements. The central measure is compliance with sustainability requirements in order to ensure regulatory conformity and avoid reputational risks. At the same time, financial stability is maintained, for example, by refraining from first mover investments in a particular non-scaled SAF technology. The focus is on the following risks and their mitigation:

- Supply risk: In principle, the current availability of SAF is limited. However, industry consensus is that there will be no major supply bottlenecks until 2030 based on the available and planned production capacities of biogenic SAF. This statement is based on an optimistic market assessment. Risks such as unsecured financing, slow approvals for production facilities and economic uncertainties remain.[27] As part of a risk mitigation strategy, airlines may also be able to stock up on the necessary quantities of SAF at short notice without running the risk of not fulfilling mandates.
- Cost risk: All airlines have to bear the higher costs of SAF compared to conventional fuel. As part of a risk mitigation strategy, contracts are therefore concluded at short notice. Airlines have to accept that there is no leeway for negotiating prices and conditions, as long as no long-term contracts are signed with producers.
- Reputational risk: The use of SAF can lead to reputational damage if there is a lack of transparency regarding the production process and supply chain or in the case of questionable raw materials (e.g., palm oil). This risk can be minimised by contracting SAF from certified raw materials and communicating the company's own activities to the general public.
- Regulatory risk: Future legal requirements could change at short notice, for example with regard to blending quotas or certificates of origin. Here too, the risk mitigation strategy can ensure that short-term room for manoeuvres is maintained by not investing in a specific production process or concluding long-term supply contracts.

Airlines that pursue a risk mitigation strategy currently consider the overall risk in connection with SAF to be low. Nevertheless, continuous monitoring of the market and the political environment remains essential.

Implementation of a Business Opportunity Strategy

The potential of SAF can be utilised as part of a "business opportunity" strategy. To this end, airlines are pursuing innovative SAF procurement management and customised monetisation approaches. This allows not only sustainability goals to be achieved, but also a competitive advantage in the evolving aviation landscape.

SAF offers environmental benefits as it significantly reduces GHG emissions compared to conventional jet fuel. However, the high cost of production is a major barrier to widespread adoption. As regulatory requirements for SAF are tightened and airlines introduce higher SAF quotas to meet own decarbonisation targets, the financial burden will increase. This will ultimately impact airline profitability and competitiveness. Many airlines are trying to pass these costs on to customers. It is critical for airlines to develop robust strategies to reduce costs and maximise revenues.

In order to overcome the challenges of SAF, airlines must develop innovative procurement strategies that go beyond traditional approaches to fuel purchasing. The contractual arrangements for the purchase of SAF are far more complex than for conventional fuel. It requires clear sustainability criteria, complete documentation, flexible pricing mechanisms and far-reaching control rights. By purchasing SAF, airlines assume joint responsibility for the entire supply chain with economic and reputational consequences.

Airlines generally calculate their fuel requirements on the basis of existing aircraft and aircraft orders, their flight efficiency and the rotation planning for operational aircraft.[28] Depending on the flight planning, weather conditions at individual airports and on certain flight routes must also be taken into account. The forecast periods range from very short-term planning to medium-term forecasts for several years.

Conventional fuel is purchased on the one hand on the basis of medium-term supply contracts up to 36 months, and on the other hand on the spot market on the basis of daily prices. A large number of airlines also use hedging instruments to protect themselves against price fluctuations. To this end, options, futures contracts and swaps are traded with the aim of achieving price and planning security.

By securing SAF early at prices below the competition, airlines can reduce the financial burden while increasing the use of SAF, creating a strong demand signal and encouraging the construction of new production facilities.

Strategic options are:

- Signing major offtake agreements: Airlines can secure long-term supply by committing to large offtake agreements that allow SAF producers to secure the bankability of projects. Long-term supply contracts with SAF producers can secure price advantages before demand and thus the price rises sharply. It is also important to define a price corridor that can be protected with options and puts to manage price volatility in the evolving market. The agreement between Air France–KLM and TotalEnergies from 2024 is an exemplary case. Over a period of 10 years, up to 1.5 million tonnes of biogenic SAF will be supplied.[29]
- Risk sharing: Market prices, political framework conditions and subsidy instruments are still unstable. As there are different pathways of SAF (HEFA, Fischer-Tropsch, PtL) and their scalability and long-term availability vary, airlines could rely on technologies that later prove to be

inefficient or not scalable. There is therefore a risk of stranded investments in supply contracts or technologies if the market develops differently than expected. By contractually sharing the multiple risks, airlines can negotiate lower SAF prices while strengthening partnerships with suppliers.

- Formation of purchasing consortia: Airlines can strengthen their negotiating position by joining forces with other purchasers to buy SAF together. Meta, JPMorgan Chase, Deloitte and AstraZeneca, among others, have joined forces in the Sustainable Aviation Buyers Alliance (SABA) to enable investment in SAF production and achieve economies of scale by pooling demand.[30]
- Utilisation of global networks: In principle, it is possible for airlines to use their global networks and book and claim systems to tap into international markets with lower SAF costs. This enables demand where no SAF is physically available. Transport and infrastructure costs are minimised. At present, the eligibility of SAF certificates for emissions targets (e.g., CORSIA, EU ETS) is not yet standardised. The EU Commission has also not yet finalised its position on this "flexibility mechanism".
- Financing production expansions: Investing in SAF production facilities or increasing the proportion of jet fuel in existing facilities can help airlines secure supply on favourable terms. British Airways, for example, is a strategic investor in the LanzaJet project in Georgia (USA), which is announced to produce 38 million litres of SAF from ethanol every year.[31]
- Equity investments: By taking equity stakes in SAF assets, airlines can negotiate even better cost-plus arrangements that ensure low-risk and predictable pricing over time. Norwegian has secured early access to production volumes by acquiring a stake in PtL producer Norsk e-Fuel in 2024.[32]

In addition to these policies, airlines operating in the US can leverage federal subsidies under the Inflation Reduction Act and the Direct Air Capture (DAC) initiative, as well as state-level incentives, to further reduce costs.

Together, these mechanisms allow airlines to source SAF at economically feasible prices while creating the confidence and risk-sharing necessary to drive production growth and facilitate the opening of new production facilities.

Procurement alone is not sufficient to realise the full potential of SAF. Airlines can use sustainability as a brand feature and thus differentiate themselves from the competition as a "green" airline. Sustainability becomes a revenue stream. A green image can strengthen the brand image and lead to a higher valuation by customers. Airlines are thus able to pass on the sustainability costs to the customer in the form of surcharges or higher prices. Airlines are obliged to invest in CO_2 emission certificates and can trade them. This gives the airline the opportunity to sell emission-relevant certificates or to generate profits from their emissions if they offset them. In this sense, airlines need to focus on monetising sustainability by tailoring their offerings to the unique motives and willingness-to-pay of different consumer segments. This includes developing customised solutions for different customer groups:

Air Cargo Customers

These customers are usually companies with their own published sustainability targets, which often adopt the roadmap outlined by the Science Based Targets Initiative (SBTi). An example of this is DHL Group, which has been officially listed in the SBTi since 2017. The company is committed to decarbonising its logistics chain, including air freight. The target is net zero emissions by 2050 with an interim target of a 50% reduction in CO_2 emissions by 2030. To this end, DHL is investing in SAF as a key measure to reduce emissions in air transport. The use of SAF is offered as part of the GoGreen Plus option in air freight, which enables customers to reduce transport-related emissions. Air freight transports goods with a high value density. For this reason, the net cost of SAF is relatively low compared to the value of the products transported. Even high SAF quotas can be realised without major price mark-ups. Based on an estimate by the consulting firm Roland Berger, a 100% introduction of SAF would lead to a

- ~ 0.5% end price increase for a laptop,
- ~ 1% end price increase for an item of clothing or
- ~ 2% end price increase for a bouquet of flowers.[33]

However, it is not enough to pass on the costs. Airlines need to communicate this benefit effectively to increase demand. Carriers can implement strategies ranging from three-way agreements, where they are compensated for SAF purchases through handling fees, to full willingness-to-pay approaches that include green offerings tailored to the specific needs of cargo customers.

Corporate Customers

Similar to cargo customers, corporate customers commonly have publicly announced sustainability targets, often with ambitious commitments for the years 2030–2040. In many industries – e.g., banking, professional services, legal, etc. – flying does account for a significant proportion of total corporate Scope 3 emissions. This means that any measure that is meaningful in the short term requires the decarbonisation of aviation, usually through SAF. One example of this is Microsoft, a major customer in the area of corporate air travel management. The company aims to be climate neutral by 2030 and climate negative by 2040. In addition, Microsoft is committed to completely eliminating all emissions caused historically since its foundation (1975) by 2050, in line with the SBTi and the GHG Protocol. By rethinking supply management, airlines can align the decarbonisation goals of these customers by creating compelling narratives that address their sustainability ambitions.

Individual Customers

Airlines often offer surcharges or voluntary carbon offsetting, where passengers can offset their carbon footprint for an additional amount (see also Chapter 9). This amount is then invested in projects that help to reduce GHG emissions, such as investments in renewable energies. Some airlines offer loyalty-based programmes where travellers receive bonus miles for environmentally conscious choices such as carbon offsetting or flying with sustainable fuels. These programmes can also be used for a fee for more miles or special sustainable benefits.

Recent findings show that customers are willing to pay under particular circumstances.[34] However, there is a behaviour-attitude discrepancy, which means that conversion rates remain low. This trend can be partly attributed to the lack of convincing offers that appeal to customers' values.[35]

The decision in favour of a particular airline is a mostly rational process for many passengers that depends on price and flight times. However, travellers show negative reactions when so-called socio-moral surcharges are levied on the airfare.[36] This can be avoided by credibly communicating to customers that the company fulfils its (ecological) responsibility and does not shift this on to the customer. Rather, it is about to justify the surcharge or the elevated price as a benefit to the customer.[37] Differentiated values play a central role here, in particular functional value, emotional value and social value.[38] These three levels are increasingly influencing consumer behaviour in the aviation industry, particularly against the backdrop of growing expectations in terms of sustainability and brand ethics. The term "value" (benefit evaluation) used here must be contrasted with "values" (norms). Such a value is generally considered before the purchase decision but is complemented by a product experience in retrospect or in the case of repetitive purchases.[39]

Functional value describes the perceived practical benefit of an offer. In the case of air travel, this primarily concerns value for money, ease of booking, punctuality and reliability as well as transparent information.

From a functional perspective, sustainable aviation is primarily about technological and operational innovations: fuel-efficient aircraft, waste avoidance in catering, paperless processes and optimised flight routes are key initiatives. Customers expect specific measures that not only achieve ecological effects but also increase comfort while travelling. The functional dimension of sustainable offers such as CO_2 transparency when booking is increasingly becoming a differentiating feature in the competitive environment. Many customers compare offers on the basis of these functional criteria not only for reasons of convenience, but also because they are seen as an expression of professionalism and customer focus.

Travelling is always associated with emotions, such as new experiences, changed perspectives and personal development. Emotional values therefore play a major role in the purchasing decision. Air travellers not only want to feel safe but also valued and understood. Brands that build a positive emotional relationship with their customers, for example through individualised customer approach, storytelling with purpose or positive experiences at contact points (e.g., app, lounge) achieve significantly higher customer loyalty.[40] Sustainability also plays a role on an emotional level. Those who fly with an airline that acts in a credibly environmentally conscious manner experience a good sense of responsibility, a kind of "moral travel experience". The feeling of being part of a positive change by choosing a responsible airline has an identity-building effect. Customer loyalty is created through the alignment of brand values with personal convictions.

More and more customers are factoring social values into their buying decisions. As global players, airlines have a particular social responsibility. Sustainability here means not only the use of low-emission aircraft or the use of SAF, but also transparent communication, fair working conditions and support for local communities. Airlines credibly assume social responsibility, for example through partnerships with environmental organisations, investments in education or the promotion of diversity, and by that create trust and gain social acceptance. Airlines that are clearly committed to social standards, inclusion, fair working conditions and environmental responsibility are favoured by many customers. Such measures have an image-building effect and position the airline as part of a responsible, modern society. The brand thus becomes part of a social change towards greater climate awareness. For many customer groups – particularly younger target groups and companies with high sustainability standards – these aspects are no longer optional but are decisive when choosing a carrier.

Using the example of a purchase option for Green Fares (ecologically sustainable airfares) it is obvious that additional measures are required to exploit the potential of monetisation (see also Chapter 9). Lufthansa Group started to introduce Green Fares on short, medium and long-haul routes in 2023. The response rate is currently around 4% and the purchase of SAF is only included proportionally.[41] The climate protection projects associated with the Green Fares have been criticised as "Greenwashing". SAS, United, JetBlue follow the model, but with limited scope. The necessity arises from the fact that the term "green fare" is viewed critically. Lufthansa has been obliged by the Dutch competition authority ACM (Autoriteit Consument & Markt) to stop labelling its "Green Fares" as "green", as this term is misleading. The background to this is that airlines in Europe are becoming more and more suspected of Greenwashing. According to the ACM, the fares suggest that booking this option makes a flight climate neutral or particularly environmentally friendly. ACM sees no factual basis and argues that the measures associated with the product, such as SAF purchase and CO_2 compensation, are not sufficient to characterise a flight as "sustainable". It is also criticised that there is no transparency as to how the emissions are calculated and offset.

Customers initially expect a functional value from the purchase of green fares. Green fares should be intuitively bookable and clearly visible as an option in the booking system. Furthermore, a green fare must not be perceived as overpriced compared to conventional tickets. Customers expect a clear explanation of what they are paying more for. The information about the ecological benefits (e.g., use of SAF) must therefore be credible and understandable. On the other hand, customers also expect that there will be no loss of comfort (e.g., choice of seat, luggage) despite a "green" decision. Many passengers are looking for ways to make their journeys more environmentally conscious. A Green Fare gives them the feeling of "doing the right thing". The purchase conveys consistency with personal values (e.g., environmental awareness, climate protection). The contribution to sustainability can be linked to a positive self-image as a conscious, modern or responsible person. The purchase of a Green Fare can be perceived as a sign of responsibility, especially in a personal environment in which a commitment to sustainability and the corresponding actions are valued. It is particularly important for companies and business travellers to use environmentally friendly options in order to maintain a positive public image. With Green Fares, passengers expect functional transparency and fairness, emotional affirmation and social recognition. For airlines, this means that they must not only implement sustainable fares technically but also communicate them in a value-orientated way and charge them emotionally so that they are more than just a "green extra".[42]

Conclusion

In summary, the combination of procurement and monetisation strategies is crucial, as aligning the characteristics of the SAF purchased with the SAF certificates sold (e.g., carbon intensity, volumes, costs) allows airlines to align incentives and expectations across the SAF supply chain. There is indeed a continuum of structures for agreements that are possible in the Business-to-Business (B2B) space (with cargo and corporate customers): from non-binding (low-risk) contracts, to "back-to-back" contracts where airlines contract separately but in close coordination with SAF suppliers and B2B customers, to three-way agreements involving the SAF supplier, the airline and the B2B customer, to full Special Purpose Vehicles dedicated to risk management.

Key Takeaways

- Sustainability has become a core strategic driver in aviation, fundamentally reshaping how airlines define long-term goals, competitive positioning and value creation.
- Strategy development in airlines must integrate external pressures such as climate regulation, stakeholder expectations and market dynamics alongside internal capabilities and resources.

- SAF represents the most relevant short- to medium-term strategic lever for reducing Scope 1 emissions while using existing aircraft and infrastructure.
- Airlines face a strategic trade-off between risk mitigation and business opportunity creation, where compliance-oriented approaches ensure legitimacy, while proactive investments can generate competitive advantages.
- Long-term strategic success requires balancing environmental ambition with economic viability, aligning sustainability initiatives with cost structures, revenue models and customer willingness-to-pay.
- Early and strategic engagement in SAF procurement and partnerships can strengthen resilience, secure future supply and position airlines as leaders in a decarbonising industry.
- Strategy development is an ongoing, dynamic process, requiring continuous adaptation as technologies, regulations and market conditions evolve.

Notes

1 Porter (1996).
2 Walls (2022).
3 IATA (2025a).
4 IATA (2025b).
5 Eurocontrol (2024).
6 GHG Protocol (2021).
7 Boeing (2025) and Airbus (2025).
8 Walls (2022).
9 Hart and Milstein (2003).
10 Hart and Dowell (2010).
11 Building (2018).
12 Changi (2025).
13 The implementation of sustainable ATM measures is associated with technological, institutional and political hurdles. The harmonisation of airspace in Europe as part of the Single European Sky (SES) has been stalled for years, as national sovereignty interests stand in the way of full integration. In addition, the digitalisation of ATM requires considerable investment in infrastructure, training and regulatory adjustments.
14 ICCT (2020).
15 Eurocontrol (2018).
16 These goals date back to 2009/2010 and were enacted in 2010 by an official resolution at the 66th IATA General Assembly. The three key goals still form the core of the international aviation industry's climate strategy today.
17 The improvement in fuel efficiency refers to the fuel consumed per passenger or tonne-kilometre for newly registered aircraft types. The ICCT (International Council on Clean Transportation) states that the average fuel consumption of new commercial aircraft was significantly improved by 2016. The target is considered to have been exceeded. Since 2020, Airbus has developed and marketed the A321XLR and the Chinese manufacturer COMAC the C919, new aircraft types that enable further efficiency gains.
18 IATA (2021) and ATAG (2021).
19 T&E (2021) and Abelvik-Lawson and Wockner (2021).

20 Álvarez and Di Nino (2017).
21 In June 2024, a test flight of an Airbus A350 aircraft with 100% SAF was successfully completed as part of the ECLIF3 study (Airbus, Rolls-Royce, German Aerospace Centre (DLR), NESTE). An official authorisation (e.g., from EASA) for the use of 100% SAF in regular flight operations is still pending.
22 IATA (2025d).
23 ICAO (2025).
24 Eccles, Ioannou and Serafeim (2014).
25 Berns et al. (2009).
26 IATA (2025c).
27 CENA (2025).
28 Excluding reserve aircraft.
29 TotalEnergies (2024).
30 Flysaba (2024).
31 LanzaJet (2021).
32 Norwegian (2024).
33 Ibarra et al. (2025).
34 Hui, Itani and O'Connell (2024).
35 The peculiarity that surveys on climate protection are influenced by socially desirable response behaviour was already discussed in Chapter 2.
36 Carattini, Kallbekken and Orlov (2019) and Levi (2021).
37 Goenka and Bagchi (2024).
38 Hur, Kim and Park (2013).
39 Boksberger and Melsen (2011).
40 Wang et al. (2025).
41 Lufthansa (2024).
42 At this point, it should be noted that studies indicate that socio-moral surcharges should be integrated into the price in the same way as Green Fares and should not be labelled as additional costs – see Goenka and Bagchi (2024).

References

Abelvik-Lawson, H, Wockner, S. (2021) Airlines are selling carbon offsets as the solution to climate change. Here's why they're wrong; https://www.greenpeace.org.uk/news/airlines-carbon-offsets-solution-climate-change-wrong/

Álvarez, I., Di Nino, V. (2017) The oil market in the age of shale oil; in: *ECB Economic Bulletin*, Issue 8; https://www.ecb.europa.eu/pub/pdf/other/ebart201708_01.en.pdf

Airbus (2025) Global Market Forecast; https://www.airbus.com/en/products-services/commercial-aircraft/global-market-forecast

ATAG – Air Transport Action Group (2021) Waypoint 2050, 2nd edition; https://aviationbenefits.org/media/167417/w2050_v2021_27sept_full.pdf

Berns, M., Townend, A., Khayat, Z., Balagopal, B., Reeves, M., Hopkins, M.S., Kruschwitz, N. (2009) The Business of Sustainability: What It Means to Managers Now; in: *MIT Sloan Management Review*, Volume 51 (1); pp. 20–26; https://sloanreview.mit.edu/article/the-business-of-sustainability-what-it-means-to-managers-now/

Boeing (2025) Commercial Market Outlook 2025–2044; https://www.boeing.com/content/dam/boeing/boeingdotcom/market/assets/downloads/2025-commercial-market-outlook.pdf

Boksberger, P.E., Melsen, L. (2011) Perceived value: a critical examination of definitions, concepts and measures for the service industry; in: *Journal of Services Marketing*, Volume 25 (3); pp. 229–240; doi:10.1108/08876041111129209

Building (2018) Oslo Airport Expands its Terminal with openBIM® and Environmentally Sustainable Design; buildingSMART International; https://www.buildingsmart.org/wp-content/uploads/2019/04/bSI_cCase_Study_2018_Avinor_Team_T.pdf

Carattini, S., Kallbekken, S., Orlov, A. (2019) How to Win Public Support for a Global Carbon Tax; in: *Nature*, Volume 565; pp. 289–291; doi:10.1038/d41586-019-00124-x

CENA – Centre of Competence for Climate, Environment and Noise Protection in Aviation (2025) CENA SAF-Outlook 2025–2030; https://www.cena-hessen.de/projekte/saf-outlook/; available in German only

Changi (2025) Encouraging Recycling in Changi Airport; https://www.changiairport.com/en/corporate/our-sustainability-efforts/environment/encouraging-recycling.html?utm_source=chatgpt.com

Eccles, R.G., Ioannou, I., Serafeim, G. (2014) The Impact of Corporate Sustainability on Organizational Processes and Performance; in: *Management Science*, Volume 60 (11); pp. 2835–2857; doi:10.1287/mnsc.2014.1984

Elkington, J. (1997) *Cannibals with Forks. The Triple Bottom Line of 21st Century Business*; Capstone, Minnesota

Eurocontrol (2018) Eurocontrol Specification for Collaborative Environmental Management (CEM); https://www.eurocontrol.int/sites/default/files/content/documents/single-sky/specifications/EUROCONTROL-SPEC-156%20CEM%20Ed.1.1.pdf

Eurocontrol (2024) Adapting Aviation to a Changing Climate; https://www.eurocontrol.int/publication/adapting-aviation-changing-climate

Flysaba (2024) SABA Announces Historic Agreements to Purchase Sustainable Aviation Fuel Certificates to Grow Investment in Clean Fuel Technologies; https://flysaba.org/2024/04/17/sustainable-aviation-buyers-alliance-announces-historic-agreements-to-purchase-sustainable-aviation-fuel-certificates-to-grow-investment-in-clean-fuel-technologies/

GHG Protocol (2021) You, too, can master value chain emissions; https://ghgprotocol.org/blog/you-too-can-master-value-chain-emissions

Goenka, S., Bagchi, R. (2024) Price Partitioning of Socio-Moral Surcharges; in: *Journal of Consumer Research*, Volume 51 (5); pp. 873–895; doi:10.1093/jcr/ucae026

Hart, S.L., Dowell, G. (2010) *A Natural-Resource-Based View of the Firm: Fifteen Years After*; in: *Journal of Management*, Volume 37 (5); pp. 1464–1479; doi:10.1177/0149206310390219

Hart, S.L., Milstein, M.B. (2003) Creating sustainable value; in: *Academy of Management Perspectives*, Volume 17 (2); pp. 56–69; doi:10.5465/ame.2003.10025194

Hui, T.H., Itani, N., O'Connell, J.F. (2024) Examining Air Travellers' Willingness to Pay for Non-voluntary Environment-related Fees: The Case of SAF Surcharge and Carbon Taxes; https://www.hos.pub/articles/hsustain3010005

Hur, W., Kim, Y., Park, K. (2013) Assessing the Effects of Perceived Value and Satisfaction on Customer Loyalty: A 'Green' Perspective; in: *Corporate Social Responsibility and Environmental Management*, Volume 20 (3); pp. 146–156; doi:10.1002/csr.1280

IATA – International Air Transport Association (2021) Net-Zero Carbon Emissions by 2050, Press Release No 66; https://www.iata.org/en/pressroom/pressroom-archive/2021-releases/2021-10-04-03/

IATA – International Air Transport Association (2025a) Global Air Passenger Demand Reaches Record High in 2024; https://www.iata.org/en/pressroom/2025-releases/2025-01-30-01/

IATA – International Air Transport Association (2025b) Global Outlook for Air Transport – Trade, AI, and the Energy Transition; https://www.iata.org/en/publications/economics/reports/global-outlook-for-air-transport-december-2025/

IATA – International Air Transport Association (2025c) Global Feedstock Assessment for SAF Production; https://www.iata.org/globalassets/iata/publications/sustainability/global-feedstock-assessment-for-saf-production-outlook-to-2050.pdf

IATA – International Air Transport Association (2025d) SAF Production Growth Rate is Slowing Down, Essential to Correct Course Ahead of e-SAF Mandates; Press Release No. 57, December 9; https://www.iata.org/en/pressroom/2025-releases/2025-12-09-04/

Ibarra, D., Ruf, Y., Dehne, F., Herr, D., Kristjansson, T., Sachdeva, N. (2025) Unlock the full potential of sustainable aviation fuels; https://www.rolandberger.com/en/Insights/Publications/Unlock-the-full-potential-of-sustainable-aviation-fuels.html

ICAO – International Civil Aviation Organization (2025) State Action Plans (SAP) and Assistance; https://www.icao.int/environmental-protection/pages/climatechange_actionplan.aspx

ICCT – The International Council on Clean Transportation (2020) CO_2 Emissions from Commercial Aviation; https://theicct.org/wp-content/uploads/2021/06/CO2-commercial-aviation-oct2020.pdf

Kumar, S., Rutherford, D. (2024) Lifetime emissions from aircraft under a net-zero carbon budget; *ICCT Working Paper*; https://theicct.org/wp-content/uploads/2024/07/ID-188-%E2%80%93-Committed-emissions_wp_final.pdf

LanzaJet (2021) LanzaJet Welcomes British Airways Aboard; https://www.lanzajet.com/news-insights/lanzajet-welcomes-british-airways-aboard-as-new-investor

Levi, S. (2021) Why Hate Carbon Taxes? Machine learning evidence on the roles of personal responsibility, trust, revenue recycling, and other factors across 23 European countries; in: *Energy Research & Social Science*, Volume 73; 101883; doi:10.1016/j.erss.2020.101883

Lufthansa (2024) Now available worldwide: more sustainable flying with Green Fares; https://newsroom.lufthansagroup.com/en/now-available-worldwide-more-sustainable-flying-with-green-fares/

Norwegian (2024) Norwegian becomes co-owner of Norsk e-Fuel; https://media.uk.norwegian.com/pressreleases/norwegian-becomes-co-owner-of-norsk-e-fuel-3297613

Porter, M.E. (1996) What Is Strategy?; in: *Harvard Business Review*, Volume 74 (6); pp. 61–78; https://hbr.org/1996/11/what-is-strategy

T&E – Transport and Environment (2021) Carbon offsets used by airlines not credible, investigation shows; https://www.transportenvironment.org/articles/carbon-offsets-used-by-airlines-not-credible-investigation-shows

TotalEnergies (2024) Air France–KLM ramps up its SAF offtake agreement with TotalEnergies, which will supply up to 1.5 million tons of more sustainable aviation fuel over a 10-year period; https://totalenergies.com/system/files/documents/totalenergies_Air-France-KLM-ramps-up-its-SAF-offtake-agreement-with-TotalEnergies_en_pdf.pdf

Walls, J.L. (2022) Introducing Sustainable Aviation Strategies; in: Walls, J.L., Wittmer, A. (Eds) *Sustainable Aviation – A Management Perspective*; pp. 91–108

Wang, Y., Jin, Z.-J., Jin, C.-H., Kan, C. (2025) Building Customer Loyalty Through Emotional Connection: How Service Provider Rapport Drives Sustainable Business; in: *Sustainability*, Volume 17 (6); 2396; doi:10.3390/su17062396

9 Sustainability Marketing in Aviation

Introduction

The aviation industry is at the centre of the global sustainability debate. On the one hand, it is a driver of economic and social connectivity, but on the other, it is a significant contributor to climate change (see Chapter 2). If other non-CO_2 effects such as contrails and nitrogen oxides are taken into account in addition to CO_2 emissions, the total climate impact of aviation is up to 3.5%.[1] In view of this, communication about environmental responsibility, company-specific objectives and measures taken by airlines and airports is gaining increasing strategic importance. At the same time, this communication currently faces a number of challenges within which Green Marketing must operate.

Key Challenges in Sustainability Communication

Challenge 1: Greenwashing and Misleading Environmental Claims

Owing to its emission-intensive nature there is a high risk that Sustainability Communication in aviation will be perceived as Greenwashing. Greenwashing is defined as the exaggerated and misleading portrayal of a company's environmental performance. The "Sins of Greenwashing" are fibbing (false statements), hidden trade-offs, no proof, vagueness, irrelevance, lesser of two evils and worshiping false labels.[2] Airlines, as the most visible actors in the aviation value chain, frequently employ sustainability-related claims such as "climate-neutral flying" or "Sustainable Aviation Fuels". Airports, on the other hand, primarily act as infrastructure providers with less visibility and responsibility for emissions. Airports mainly cause indirect or secondary emissions (e.g., electricity consumption, ground traffic). Greenwashing occurs where brands communicate. Airlines are in direct contact with consumers (B2C), whereas airports mainly operate in the B2B sector. The less contact there is with end consumers, the lower the risk. Airports are often subject to local environmental regulations (noise protection, energy efficiency) and economy-wide CO_2 reduction targets on the national (or

DOI: 10.4324/9781003608172-9

European) level. The Greenwashing debate therefore focuses on airlines whose communication strategies raise immediate ethical and reputational issues. Greenwashing allegations can have negative effects for airlines and undermine brand trust in the long term.[3]

Airlines are subject to international climate protection initiatives (e.g., CORSIA, EU ETS) and are increasingly in the political and regulatory focus. A study from 2024 examines the effectiveness of the EU Greenwashing Directive (EU 2024/825) using the example of voluntary CO_2 compensation programmes by European airlines. The study shows that almost all airlines meet these criteria at least partially, albeit most only to a limited extent. Many programmes are largely symbolic in nature and have little measurable impact on reducing emissions. Airlines use their own calculation models, questionable carbon credits and non-standardised sustainability indicators, which give consumers an overly positive picture.[4] Although the EU directive improves transparency requirements, it remains insufficient as it does not require uniform measurement standards or binding proof of effectiveness. The traceability and actual climate impact of compensation measures vary considerably. It often remains unclear whether the projects actually result in a permanent reduction. Another point of criticism is that compensation often serves as a substitute for actual emission avoidance ("compensation instead of reduction"). Consequently, it can be concluded that CO_2 offsetting serves more to ease consumers' consciences than to bring about substantial reductions in emissions. Greenwashing can significantly undermine consumer confidence. When customers perceive Greenwashing, it weakens their trust in companies and negatively influences their purchasing behaviour. Information asymmetries further exacerbate this effect. Complex technical relationships and unclear sustainability indicators often contribute to information deficits. The result of such communication shortfalls is a loss of credibility. When sustainability promises are perceived as exaggerated or marketing driven, the confidence of consumers, investors and regulatory authorities declines.[5]

For example, advertising campaigns by major airlines such as Air France, Lufthansa and Etihad were banned by the British Advertising Standards Authority for making misleading environmental claims.[6] The ruling by an Amsterdam district court against KLM marks another precedent in Europe's approach to Greenwashing in the aviation industry. The court found that KLM's advertising for "sustainable flying" contained misleading and unsubstantiated environmental claims. This applies in particular to Sustainable Aviation Fuel (SAF) and carbon offset programmes. Although no recall obligation was imposed for previous advertising campaigns, the ruling obliges KLM to communicate honestly and specifically about climate measures in future.[7] The case highlights the growing regulatory control over sustainability statements. Another recent example of this development is provided by the European Commission's press release published in November 2025. Following a dialogue with the Commission and national consumer authorities,

21 European airlines agreed to revise their environmental communication.[8] They committed to refraining from making misleading statements about supposedly "climate-neutral" flights, to only using the term "Sustainable Aviation Fuels" with scientifically sound evidence and to providing valid information about actual emissions and planned reduction measures in future. In addition, future statements on "net zero" targets must be backed up by clear timetables and verifiable intermediate targets.[9] These measures underscore the growing importance of verifiability in Sustainability Communications and show that Greenwashing is no longer considered merely a reputational risk, but rather legally relevant market behaviour. In this context, EU Directive 2024/825 against misleading environmental advertising is becoming increasingly important, as it creates the basis for uniform standards of environmental communication in emission-intensive industries.[10]

Challenge 2: Complexity and Scientific Uncertainty

Sustainability in aviation is embedded in a highly complex technological and regulatory environment and is torn between the demand for mobility and climate protection. The industry is characterised by long investment cycles and infrastructural constraints, which make a fast transformation difficult. The sector is hard to decarbonise because aircraft have long life cycles and alternative propulsion technologies (e.g., electric or hydrogen aircraft) remain in early development stages.[11] Moreover, regulatory frameworks such as CORSIA and the EU ETS evolve continuously, and global fuel markets impose structural constraints. These characteristics impede rapid decarbonisation and complicate the communication of credible progress. Sustainability Marketing in aviation must therefore address not only communicative but also structural barriers.

In addition, scientific uncertainties challenge the accuracy and comparability of sustainability information particularly with regard to non-CO_2 effects such as contrails. In public communication, these aspects are often only mentioned in passing or ignored altogether, which can create a distorted picture of the actual ecological challenge. Airlines apply divergent system boundaries and employ different calculations (e.g., emissions per seat kilometre, per passenger kilometre or absolute emissions). These inconsistencies hinder stakeholder comprehension and impair the comparability of sustainability statements.[12] As these methodological limitations restrict the communicative space of airlines, they highlight the need for a systematic marketing approach that integrates complexity responsibly and clearly. This role can be fulfilled by Sustainability Marketing.

Challenge 3: Lack of Transparency and Greenhushing

Despite the increasing prevalence of sustainability reports, it often remains unclear what criteria are used to assess environmental performance. Many airlines publish figures, but do not explain transparently how these were collected or what limitations the data compilation has.[13] In practice strategic long-term goals such as "Net Zero 2050" are often communicated without providing a clear explanation of interim targets, concrete actions or measurable milestones. This creates a narrative of aspiration but no reliable report on progress.[14]

At the same time, a countervailing trend can be observed: so-called Greenhushing, i.e. the (in some cases) deliberate withholding of information about sustainability activities for fear of criticism or accusations of Greenwashing. While most European airlines now routinely report basic emissions data such as total CO_2 emissions, they tend to fall short in reporting their actual sustainability achievements. One example of this is the Lufthansa Group, which regularly publishes detailed data on its Scope 1 CO_2 emissions in accordance with regulatory requirements. However, the airline provides comparatively limited visibility of the concrete scale and impact of its sustainability achievements, particularly investments in SAF. For instance, while Lufthansa has entered multiple SAF procurement and production partnerships, such as its agreement with HCS Group, the actual volumes currently deployed represent only a small fraction of the company's total annual fuel consumption. This discrepancy between transparent emissions reporting and the undercommunication of early-stage SAF implementation can lead stakeholders to underestimate both Lufthansa's efforts and the systemic challenges associated with large-scale SAF adoption. Achievements such as long-term procurement agreements, early-stage production partnerships, or modest initial blending volumes, have comparatively limited transparency. Such a communication gap can create a distorted perception of environmental performance, as quantitative emissions figures are highly visible. As a result, stakeholders may misjudge both the strategic relevance and the practical challenges associated with scaling SAF, thereby hindering informed evaluation of airlines' transitional pathways.

These dynamics demonstrate the need for communication strategies that balance accuracy, transparency and contextualisation.[15] Sustainability Marketing addresses this need by establishing structured principles for credible environmental communication.

Challenge 4: Conflicting Objectives and Limited Comparability

Sustainability in aviation is caught between ecological ambitions and economic burdens. Airlines operate in a highly competitive global market where cost efficiency and growth remain key success factors. According to the European Aviation Safety Authority (EASA), the main barriers to more sustainable aviation are the limited availability of SAF, high production costs and inadequate infrastructure.[16] Nevertheless, many companies are focusing their communications on technical progress and transformation, while at the same time expanding their fleets and route networks. This contradiction is often criticised by environmental organisations.[17]

Furthermore, the lack of standardisation in reporting makes an objective assessment difficult. Without uniform framework conditions, such as those required by the Global Reporting Initiative (GRI), Sustainability Communication often has a limited value to stakeholders.[18] Such contradictions underscore the necessity for marketing concepts that integrate economic, regulatory and ecological objectives. Sustainability Marketing offers a framework that allows companies to position sustainability not merely as a compliance requirement but as a strategic part of value creation and brand positioning.

Challenge 5: Limited Target Group Orientation and Comprehensibility

Sustainability Communication often fails to account for differing levels of expertise among stakeholders. Technical language, complex indicators and abstract concepts (e.g., radiative forcing (RF), life cycle assessment) limit the comprehensibility of sustainability reports for general audiences. These concepts are difficult for the general public to understand, which limits the effectiveness of communication. While regulators and investors are familiar with technical indicators, customers and the media often fail to see the connection to specific areas of action and impact. One example of this is communication about the use of SAF. Airlines often emphasise their percentage blend or CO_2 reduction over the entire life cycle without explaining what this means in practice. Terms such as "drop-in fuels" or "feedstock diversification" remain incomprehensible to many.[19] Effective Sustainability Marketing requires messages and formats that reflect the distinct expectations, knowledge levels and decision-making logics of specific audiences:

- Passengers primarily respond to simple, emotionally resonant and benefit-oriented information. Complex technical details often overwhelm this group; instead, intuitive explanations of tangible improvements such as less noisy aircraft are more effective.
- The public and media require clear, consistent and verifiable claims. Ambiguous terms like "climate-neutral" increase reputational risks. Communication for this audience must therefore emphasise methodological transparency and avoid overstating environmental achievements.
- For policymakers and regulators, accuracy and compliance are central. This group expects transparent reporting, clear reduction pathways and honest communication about uncertainties and limitations.
- Investors evaluate sustainability in terms of risk, future viability and credible transition strategies. They require quantitative indicators on fleet renewal, SAF procurement and long-term decarbonisation plans.
- Finally, employees rely on sustainability narratives for identity-building and organisational cohesion. Internal communication that links environmental goals to everyday work strengthens motivation and employer attractiveness.

Taken together, the challenges outlined above reveal that Sustainability Communication in aviation is facing a dilemma. It is vital for establishing transparency, accountability and social legitimacy, but at the same time, it is impeded by structural contradictions, technological uncertainties and conflicting economic objectives. The key challenges of Sustainability Marketing relate to avoiding Greenwashing, improving data and methodological transparency, unifying reporting standards and developing comprehensible communication formats. These challenges make it clear that sustainability in aviation is not just a technical or regulatory issue, but above all a question of strategic market development. Only when airlines and airports communicate honestly about the progress and limitations of their sustainability strategies is trust in the ecological transformation of aviation possible. The credibility of Sustainability Communication is not a by-product of PR strategies, but a question of corporate integrity and the consistent scientific basis of statements. Sustainability Marketing responds to precisely these requirements. By situating sustainability within the established framework of marketing it provides a systematic method for translating environmental objectives into market-oriented strategies and stakeholder-specific communication. The next section therefore builds on the challenges identified and elaborates how the Marketing Mix can be designed to enhance credibility, strengthen stakeholder confidence and support the ecological transformation of the aviation sector.

Sustainability Marketing in Aviation

Fundamentals of Sustainability Marketing in Aviation

Sustainability Marketing is a key component of modern corporate strategies in the aviation industry. It aims to combine environmental and social responsibility with market- and profit-oriented activities.[20] In aviation, Sustainability Marketing refers to the systematic incorporation of ecological and social principles into product design, market communications and corporate positioning. Its purpose is not only to minimise the environmental footprint of aviation services but also to strengthen competitiveness, trust and legitimacy in a sector under intense public scrutiny.[21] Beyond image building, Sustainability Marketing in aviation has a transformative function. It shapes consumer expectations, influences industry structures and contributes to a shift towards more sustainable value chains. This broader responsibility includes a wide range of stakeholders: passengers and freight clients, political and social actors, investors and financial institutions, suppliers, non-governmental organisations (NGOs) and employees.[22] It not only communicates responsibility but also serves as a vehicle for strengthening one's brand in the long term. Sustainability Marketing serves to cultivate one's image, but increasingly also to legitimise economic activity.[23] In view of the high level of public pressure, climate policy regulations and social awareness of environmental issues, it is essential for Airports, Airlines, Air Navigation Service Providers (ANSP) and Ground Handling Companies to position themselves credibly as sustainable players.

The following sections outline the core functions of Sustainability Marketing and examine its operationalisation through the Green Marketing Mix, complemented by behavioural economic instruments that influence consumer decision-making.

Core Functions of Sustainability Marketing

Differentiation and Image Building

Air transport is highly competitive, price sensitive and characterised by low inherent service differentiation. Since the core product (transporting passengers and goods) is functionally homogeneous, sustainability enables the creation of emotional and value-based added benefits and can become a strategic differentiator.[24] Examples include investments in early-stage SAF projects, transparent emissions reporting at the point of sale, CO_2 compensation products and social responsibility measures such as community development projects linked to aviation operations. Sustainability-based differentiation operates on two levels. Externally, it enhances brand perception among consumers, investors and society. Internally, it fosters corporate identity and strengthens employee commitment.[25]

Empirical research shows that a credible sustainability profile increases customer loyalty and may raise willingness to pay within environmentally conscious segments. Thus, Sustainability Marketing contributes not only to reputation but to long-term brand equity.[26]

Trust and Credibility

The success of Sustainability Marketing depends on the credibility of the measures communicated. Trust is therefore not created by symbolic gestures, but by comprehensible, verifiable and long-term strategies.

Companies that report openly on their emissions and challenges and demonstrate measurable progress create a higher level of credibility. Certification schemes (e.g., ISO 14001), compliance with international reporting standards and independent verification can also enhance credibility.

Communication also plays a crucial role. Sustainability messages must be realistic, specific, data-based and formulated on an equal footing with consumers. Exaggerated or unsubstantiated promises quickly lead to a loss of trust and public criticism, which can cause considerable damage to a company's reputation in the age of digital transparency. Sustainability Marketing in aviation provides important impetus for innovation, transparency and consumer awareness, but can only be considered credible if it is based on a demonstrable reduction in environmental impact. The alignment of communicated values with operational practices remains the core driver of credibility. Consistency does not come about through communication alone, but through the integration of ecological and social principles into the entire value chain.[27] From an analytical perspective, credibility reduces information asymmetries and mitigates stakeholder scepticism. In a sector frequently criticised for its environmental impact, consistent reporting and verifiable reductions in emissions are essential to maintain a social licence to operate. In this sense, Sustainability Marketing fulfils a dual function: it sensitises consumers to the ecological impact of their mobility and at the same time forces companies to strategically integrate sustainability into their business models.

Green Marketing Mix

In order to implement this strategic integration effectively, ecological principles must be consistently anchored in all elements of the marketing activities. It means treating socio-ecological problems as a starting point of the marketing process, not as a set of externalities or constraints. Green Marketing promotes environmentally friendly processes and products, while also meeting customer expectations.[28] It consists of a wide range of reconfigured marketing activities, the so-called Green Marketing Mix. The Marketing Mix is one of the basic concepts of conventional marketing and comprises the 4 Ps: Product, Price, Placement and Promotion (see Figure 9.1).[29] By adapting these classic marketing instruments to ecological and social requirements, aviation companies can not only improve their environmental balance sheet, but also promote trust, competitiveness and long-term customer loyalty.

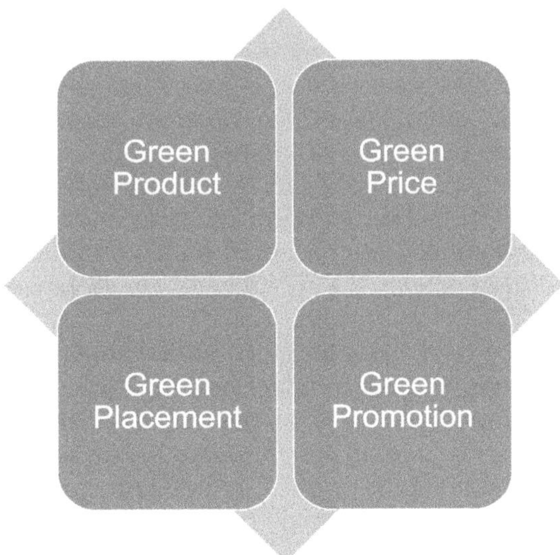

Figure 9.1. Green Marketing Mix in aviation.

Green Product

Product policy covers all areas of business activity relating to the market-oriented design of products. The basic idea is that the company's range of services must be aligned with long-term customer satisfaction.[30] In addition to these economic goals, which make a product competitive and create customer value, sustainable product policy requires consideration of environmental necessities. Green products are designed to protect the natural environment by conserving resources and reducing or completely avoiding the use of toxic substances, pollution and waste.[31] To increase clarity, green product initiatives in aviation can be grouped into four categories:

- Fleet and technology: Fleet modernisation is one of the industry's most effective environmental strategies. New aircraft types such as the Airbus A350 or Boeing 787 achieve fuel savings of up to 25% compared to older models due to advanced aerodynamics, lightweight composite structures and efficient engines. Since fuel accounts for 20–40% of airlines' total operating costs, such assets generate both ecological and economic benefits. The use of fuel-efficient aircraft is in the airlines' own economic interest, as it generates a "double dividend": on the one hand, operational savings through reduced kerosene consumption and, on the other hand, ecological advantages through lower CO_2 emissions. They also facilitate compliance with regulatory frameworks and emissions targets. Digitalisation further enhances environmental performance through data-based route optimisation, paperless cockpit procedures and predictive maintenance.
- Airports and infrastructure: Airports increasingly implement carbon-neutral operational concepts. E.g., Munich and Zurich use photovoltaic systems, geothermal energy and district heating to meet their energy needs from renewable sources.[32] Energy-efficient terminal designs, smart lighting systems and recycled construction materials contribute to reduced life cycle emissions. In addition, airports are increasingly offering green mobility solutions. Electric bus systems, bicycle rental stations and connections to public transport promote environmentally friendly arrivals and departures. The concept of "Green Airport Cities" combines operational sustainability with urban development and social responsibility.[33]
- Ground handling and operations: Ground handling companies contribute significantly to green product development. The introduction of fully electric ground vehicles, sustainable de-icing procedures using closed-loop systems and digital resource optimisation tools reduces emissions and operational waste. In addition, ground handlers are developing digital tools for resource optimisation, such as efficient resource planning and reduction of empty runs. These measures combine technological innovation with process efficiency.

- Passenger experience and ancillary products: Another area of sustainable product development concerns on-board equipment and consumables. Airlines are increasingly turning to materials that are reusable, biodegradable or sourced from recycled materials. For example, disposable items such as cutlery, crockery and drinking cups are being replaced by alternatives made from bio-based materials. Similarly, textiles and seat covers made from recycled PET bottles or biodegradable fibres are being used in modern aircraft cabins. Digital innovations such as e-boarding passes, electronic baggage tags and paperless processes are promoting the reduction of disposable materials throughout the entire travel chain.

From an analytical standpoint, green products strengthen both environmental effectiveness and customer trust, provided that product characteristics are transparent and verified. Poorly defined or misleading green products risk consumer backlash, highlighting the importance of clear value propositions. Trust-building characteristics are particularly important in this context. These must be ensured through brands or quality labels. Many green products fail in the market due to the gap between consumer expectations and perceptions. To overcome this, it is essential to find out which product characteristics are important to consumers.[34] Studies confirm that in the case of CO_2 compensation programmes, characteristics such as implementation by non-profit organisations have a positive influence on consumers' decisions to actually make a compensation payment.[35] Consumers are more willing to pay for certified carbon offset programmes run by non-profit organisations, which increases their credibility and perceived effectiveness. Customers also prefer higher levels of carbon offsetting. For example, they are more likely to choose programmes that offer a 25% reduction than those offering only 5%. In terms of programme type, most existing studies show a preference for forestry and afforestation projects.[36]

Green products comprise specific, environmentally friendly products, technologies or services that contribute directly to reducing environmental impact while strengthening customer confidence. In contrast, Green Pricing describes an economic approach that promotes sustainability through pricing and customer participation.

Green Price

Products and services in many industries have an impact on the environment through their emissions. Environmental costs are generally not included in their prices. In other words, society subsidises environmental externalities.[37] Green Pricing involves integrating environmental costs or investments in climate protection measures into the price structure. Typical examples include ticket surcharges to finance sustainable fuels or fare models that incentivise the use of environmentally friendly options. Green Pricing aims to make sustainability marketable and actively involve passengers in financing climate-friendly measures. Pricing instruments make sustainability tangible and assign economic value to environmental responsibility.

Green products cost more, which often conflicts with consumers' willingness to pay and often leads to a gap between attitude and actual purchasing behaviour.[38] Companies must convince consumers that by paying the extra cost, they are benefiting themselves, future generations or the environment. At the same time, the Green Price must be reasonable and competitive.[39]

Willingness to pay is a measure that defines the maximum price consumers accept to pay for a product or service. Studies on consumers' willingness to pay for climate protection in aviation show a wide range of amounts.[40] Willingness to pay varies strongly across consumer groups, project types and travel contexts. The price has a significant influence on air travellers' decision to pay compensation as consumers tend to choose a cheaper option.[41] Green Pricing helps internalise externalities but requires careful design. To be accepted, surcharges must be transparent, comprehensible and perceived as fair. Airlines can improve acceptance through clear explanations of fund allocation, additional personal benefits (e.g., status miles) and streamlined payment processes.[42] From a market-segmentation perspective, significant differences exist between private and business travellers. Private travellers typically operate under stricter budget constraints and display higher price sensitivity, making their willingness to pay dependent on factors such as trip purpose, household income and perceived personal benefit of climate contributions. Their decisions are often discretionary and emotionally influenced, resulting in comparatively lower and more volatile payment intentions. Business travellers, by contrast, tend to exhibit a higher willingness to pay due to corporate travel budgets, reduced personal cost sensitivity and increasing organisational commitments to sustainability targets. For many firms, climate-related surcharges can be integrated into corporate Social Responsibility strategies or mandatory internal emissions policies, which structurally elevates the acceptance of green fees. Thus, divergent economic incentives and decision-making logics between these two groups produce distinct behavioural patterns that must be considered when designing effective and targeted Green Pricing mechanisms.[43]

Green Placement

The term Green Placement refers to the complex series of decisions and activities aimed at distributing green products between their place of origin and their place of consumption.[44] From an airline perspective Green Placement refers to the environmentally responsible design of distribution channels, service delivery environments and customer touchpoints along the entire travel chain. While traditional placement in aviation focuses on the accessibility of flight services through digital booking systems, sales partners and intermodal connections, the sustainability-oriented interpretation extends this perspective to include ecological criteria in the physical and digital provision of airline services. As airlines increasingly operationalise sustainability strategies, the "greening" of distribution-related processes becomes a relevant element of credible environmental marketing.

A dominant dimension is intermodal integration, for example cooperation with rail operators, urban transport systems and mobility service providers to create low-carbon intermodal connections. Rail-air partnerships, integrated digital ticketing and the promotion of public transport for passengers and crew reduce emissions and increase the ecological compatibility of distribution routes. Such intermodal placement strategies help airlines align customer accessibility with broader decarbonisation goals.[45]

The physical placement of operational facilities is also increasingly guided by sustainability considerations, such as proximity to renewable energy, electrified ground infrastructure and circular waste systems. Furthermore, digital sales platforms influence consumer choices by providing transparent emissions information at the point of purchase.

In other words, Green Placement aligns market accessibility with decarbonisation goals and reduces life cycle emissions across the aviation value chain.

Green Promotion

Green advertising is the most important element of the Marketing Mix. It encompasses advertising and communication tools that aim to highlight the ecological benefits of the product, promote an environmentally friendly lifestyle or emphasise the company's environmental credentials.[46] Since consumers seldom actively seek green products, repeated yet meaningful communication is essential. In this context, consumers value information that is complete, easy to understand and not time consuming. Statements must be meaningful and backed up by genuine company activities. However, the information provided by airlines is often perceived as confusing, untrustworthy and lacking transparency. Claims that are difficult to understand and relate to rather subjective and ambiguous concepts (e.g., "sustainable", "communities") are combined with trustworthy statements that are verifiable (e.g., "certified").[47]

Credible and effective environmental communication by airlines requires transparency, verifiability and linguistic clarity. Companies should provide accurate and provable information in their environmental advertising, based on specific data and scientifically sound calculations. This includes, in particular, disclosure of emission levels, calculation methods and the specific projects to which compensation payments are directed. General or emotionally charged terms such as "climate neutral" or "sustainable travel" should be avoided unless they are clearly defined and proven. An essential part of trustworthy communication is the clear distinction between emission avoidance and offsetting. While technical innovations and operational efficiency measures aim at actual reductions, offsetting programmes serve exclusively to compensate for remaining emissions. A transparent presentation of this difference prevents misleading ideas about the environmental impact of flying. Furthermore, airlines should not shift responsibility onto consumers. Environmental communication must clearly state what measures the company itself is taking to reduce emissions, rather than solely emphasising the possibility of individual compensation.

Effective environmental communication is characterised by simple, understandable language and a clear structure. Specific and relevant information must be provided at the point of purchase. Airlines can make CO_2 offsets more visible to consumers by providing relevant information in newsletters, on websites and in in-flight magazines, for example. Integrating compensation offers into the booking process before the ticket is purchased can increase the perception of integrity, provided that sufficient information is available. Convenience plays an important role in this respect. These options should therefore be available during the online booking process, before the booking is completed and together with various payment options. Finally, clear standards and external verification mechanisms are needed to ensure the quality and truthfulness of environmental claims. Uniform guidelines, independent certification and regular reviews can help to avoid the problem of Greenwashing, increase comparability between providers and promote long-term trust in the aviation industry's environmental communication.

Promoting Sustainable Aviation: Oscar the Grouch and Others

In the context of Green Communication, United Airlines uses the character "Oscar the Grouch" to make sustainability issues facing the industry understandable to a broad audience while anticipating critical perceptions. In 2023, United officially appointed Oscar as "Chief Trash Officer" to launch an educational campaign on SAF, in which he explains how waste can be converted into more climate-friendly aviation fuels. Oscar has appeared in over 30 video, digital and social media formats to translate the technical content into simple, narrative messages, thereby increasing the attention of different target groups. The communicative approach was characterised by storytelling, humour and persona-based communication, whereby United Airlines presents ecological measures not only as a corporate goal, but also as a learnable process. This form of communication is intended not only to inform, but also to create emotional resonance and address common reservations before they arise as criticism. By using a character from childhood culture, United de-dramatises complex content and makes it accessible to different age groups. This Green Communication approach aims to increase credibility and convey transparent, reflective messages rather than simple PR promises. Such humorous elements can promote the perception of brand warmth and engagement, making Sustainability Communication in aviation both more memorable and emotionally relevant. The approach of communicating the topic of SAF to the public and raising awareness of more environmentally friendly aviation has been taken up and further developed by several airlines. Singapore Airlines is part of the Green Fuel Forward campaign to raise awareness of SAF in the Asia-Pacific region. Japan Airlines, among others, has been running a SAF awareness Promotion campaign in cooperation with Tokyo Narita Airport since November 2025.[48]

Behavioural Economic Instruments in Sustainability Marketing

Beyond the four marketing instruments, the use of behavioural tools such as nudging, defaults, social norms and feedback is gaining increasing importance in the context of Sustainability Marketing. These approaches aim to promote environmentally conscious behaviour not only through rational arguments or classic marketing measures, but also through the subtle design of decision-making situations. For example, preset sustainable options (defaults) or the emphasis on social norms can effectively influence sustainable consumer behaviour. Additional feedback on individual or collective environmental impacts can also raise awareness of the consequences of one's own actions and lead to greater identification with sustainable offerings in the long term.

Nudging

A nudge is a concept from behavioural economics that offers an alternative approach to communicating and promoting environmentally friendly products. While pro-environmental marketing relies on more explicit strategies, nudging is more implicit. Behavioural economics attempts to explain individuals' decision-making based on traditional economic and psychological assumptions. It assumes that people act with limited rationality and often do not make the best decisions due to a lack of information, attention and self-control.[49] The term "nudge" refers to the goal of predictably steering people's behaviour without prohibiting options or significantly altering their economic incentives. Nudging is thus based on the idea of libertarian paternalism. This aims to influence people's behaviour while preserving their freedom of choice.[50] However, a recent study shows that neither positive nor negative goal framing significantly influences travel behaviour. Goal framing means displaying CO_2 emissions in either a positive frame ("Taking the train saves CO_2") or a negative frame ("Flying produces CO_2"). Decisions about which transport option to choose are still primarily determined by price and travel time, not by the knowledge about CO_2 emissions. Travellers with a strong environmental awareness, previous experience of train travel and a social environment that supports sustainable behaviour are more likely to choose the train. Indeed, informative initiatives can raise awareness, but they are not enough to overcome deeply ingrained habits and preferences for convenience.[51]

Defaults

Default settings are settings that apply when people do not take active steps to change them. People tend to stick to the defaults for several reasons:

- Suggestion and endorsement: People believe that a default option has been pre-selected for good reason,
- Inertia: People avoid cognitively demanding thinking and default options require the least effort,
- Loss aversion: People view rejecting default options as a loss, which is felt more intensely than a gain.

Standard options can be offered in the form of opt-in and opt-out decisions. In the former case, a person must select the option, while in the latter case, the options are preselected and require active action to reject them.[52] Research shows that default options can be an effective tool for promoting carbon offsetting in aviation. Trust acts as a moderator in this context. On the one hand, consumers are more inclined to pay compensation if this measure is supported by others. On the other hand, airlines that offer such compensation payments are regarded as trustworthy if they are endorsed by society.[53] On the other side it should be noted that defaults are less effective when

- people have strong preferences or extensive knowledge (e.g., experts),
- the default clearly conflicts with existing interests (e.g., in case of excessive costs),
- there is a lack of trust in the company.

In these cases, active choosing or personalised defaults are more suitable.

However, it should be noted that opt-out default options for additional payments are currently prohibited in the EU under the Consumer Rights Directive (2011/83/EU).

Social Norms

Social norms are unwritten rules that are upheld by social approval or disapproval. These can influence individuals' behaviour for various reasons. Individuals may try to conform, avoid rejection, improve their social status, or view the behaviour of others as an indication of what is most effective. Social norms can be identified through observation or communicated directly to consumers. For example, companies may use information such as "9 out of 10 customers choose this option" or "your consumption is 10% above average". Scientific literature distinguishes between descriptive (what most people do) and injunctive (what most people approve of) social norms. Alternatively, social norms can be prescriptive (what other people do or approve of) and proscriptive (what other people do not do or disapprove of).

Numerous studies show that social norms can be an effective tool for promoting environmentally friendly behaviour. For example, social norms have a positive influence on consumers' willingness to pay for CO_2 offsets in air travel. Trust acts as a moderator in this context. On the one hand, consumers are more inclined to pay compensation if this measure is supported by others. On the other hand, airlines that offer such compensation payments are regarded as trustworthy if they are endorsed by society.[54]

Feedback

Feedback refers to information provided to consumers about the consequences of their consumption behaviour through better information and disclosure. It can take the form of personalised feedback, such as displaying water consumption in showers to encourage water saving, or informative feedback that communicates specific product characteristics (e.g., eco-labels) and helps consumers choose between different options.[55] When it comes to simple decisions, such as choosing a flavour of ice cream, consumers can easily predict what their consumer experience will be like. However, in more complex decision-making situations, such as choosing a sustainable form of travel, inexperienced and limitedly informed consumers find it difficult to assess the consequences of their actions. To overcome this, companies can present information in a more understandable and context-related way by translating various aspects into terms that are intuitively comprehensible to consumers.[56] The manner and timing of how sustainability information is presented plays a key role in its effectiveness. Information should be provided "just in time", i.e., at the very moment when consumers are making a decision. Eco-labels have a positive influence on consumers' willingness to pay, while emissions data is particularly effective when it is converted into understandable terms. For instance, instead of simply stating that a product emits 5 kg of CO_2, translating this into relatable terms such as "equivalent to running a washing machine five times" makes the information more tangible and impactful for consumers. As labels are easier to interpret and signal trust, they generally prove to be more effective than purely numerical emissions data. In addition, feedback and visual communication (e.g., via apps or displays) help consumers better understand the connection between their decisions and their environmental impact. Such feedback, whether through digital platforms, environmental labels or emotional campaigns, reduces cognitive barriers and promotes sustainable behaviour.[57]

Sustainability Marketing in aviation operates in a field of tension between ecological ambition, regulatory requirements, economic pressure and public scrutiny. Credibility cannot be achieved through communication alone but requires verifiable environmental performance and transparent reporting. By integrating sustainability into all elements of the Marketing Mix and complementing them with behavioural economic instruments, airlines and airports can strengthen trust, foster innovation and contribute to a more sustainable mobility system.

Key Takeaways

- Sustainability Communication has become a strategic necessity in aviation, but its effectiveness is fundamentally constrained by Greenwashing risks, scientific uncertainty and fragmented reporting standards, which undermine stakeholder trust if not addressed systematically.
- Greenwashing in aviation is no longer merely a reputational issue but an increasingly regulated market behaviour, as evidenced by legal rulings and EU-level directives that require verifiable, specific and methodologically sound environmental claims.
- Scientific complexity, particularly regarding non-CO_2 effects and divergent calculation methods, limits the comparability and comprehensibility of sustainability information, highlighting the need for transparent system boundaries and consistent metrics in Sustainability Marketing.
- The coexistence of Greenwashing and Greenhushing reflects a structural communication dilemma, where both exaggerated claims and the under-communication of early-stage sustainability efforts can distort stakeholder perceptions and impede informed evaluation of transition pathways.
- Sustainability Marketing provides a structured framework to reconcile ecological objectives with market-oriented strategies, embedding environmental and social principles into the Green Marketing Mix across Product Design, Pricing, Placement and Promotion.
- Behavioural economic instruments such as nudging, defaults, social norms and feedback can complement traditional marketing tools, but their effectiveness depends on trust, contextual relevance and regulatory constraints, particularly in price-sensitive aviation markets.
- Credible Sustainability Marketing in aviation ultimately requires alignment between communication and operational performance, as long-term trust and social legitimacy can only be achieved through verifiable emissions reductions, transparent reporting and consistent integration of sustainability across the entire value chain.

Notes

1. ATAG (2024) and Sharma (2025).
2. Peattie and Crane (2005), Neureiter et al. (2024), Bernini and La Rosa (2024), de Freitas Netto et al. (2020) and Delmas and Burbano (2011).
3. Bladt, van Capelleveen and Yazan (2023).
4. de Mello and Macario (2024).
5. Tsoi and Liu (2025), Hung and Chang (2023), Tu et al. (2024), Xu et al. (2025), Hyams and Fawcett (2013) and Karhunmaa et al. (2023).
6. The Guardian (2023).
7. Corder (2024).
8. Among others Lufthansa, Air France–KLM, Ryanair and SAS.
9. EC (2025).
10. EU (2024).

11 EASA (2025).
12 Iken and Aguessy (2002) and Zou, Lee and Waguespack (2025).
13 Thepchalerm and Pinsuwan (2025).
14 ATAG (2025).
15 Font, Elgammal and Lamond (2017) and Martin-Domingo, Efthymiou and Mota (2025).
16 EASA (2025).
17 T&E (2025).
18 Gudmundsdottir and Sigurjonsson (2024) and GRI/SASB (2021).
19 Rogachuk et al. (2025), Tsoi and Liu (2025), Cocolas et al. (2024), Schomakers, Engelmann and Ziefle (2024).
20 Peattie and Crane (2005).
21 Peattie and Belz (2010).
22 Groening, Sarkis and Zhu (2018).
23 Kotler, Armstrong and Opresnik (2024).
24 Leonidou, Katsikeas and Morgan (2013) and Belz and Peattie (2012).
25 Ottman (2011).
26 Hartmann and Apaolaza-Ibáñez (2012).
27 Becker-Olsen, Cudmore and Hill (2006).
28 Dangelico and Vocalelli (2017), Groening, Sarkis and Zhu (2018) and Peattie and Belz (2010).
29 Reddy et al. (2023).
30 Wirtz (2024).
31 Ottman (2011).
32 Zurich Airport (2025a), Zurich Airport (2025b) and Munich Airport (2025).
33 Oslo Airport is a prime example of this, see Life in Norway (2025).
34 Tseng and Hung (2013).
35 Li, Beck and Merkert (2026).
36 de Mello and Macario (2024) and Ritchie, Kemperman and Dolnicar (2021).
37 Arthur C. Pigou has developed the concept of externalities and uses their existence as a justification for government intervention (Pigou 1932).
38 Vieira, Castro and Souza (2023) and Gössling and Dolnicar (2022).
39 Kurek and Górowski (2025).
40 Wendt, Kipperberg and Lindhjem (2024).
41 Kortsch and Händeler (2024).
42 Cordes, Baumeister and Käyrä (2024).
43 Crosby, Thompson and Best (2024) and Buchmann (2025).
44 Sohail (2017).
45 Dobruszkes (2025), Babić et al. (2022), Hensher and Nelson (2025) and Mahesh and Calvert (2025).
46 Davari and Strutton (2014).
47 Babakhani, Ritchie and Dolnicar (2017) and Dangelico and Vocalelli (2017).
48 United (2023), Singapore Airlines (2025) and Cosmo (2025).
49 Eslaminassab and Ehmer (2021).
50 Thaler and Sunstein (2009).
51 Hollenstein and Wittmer (2024).
52 Eslaminassab and Ehmer (2021) and Sunstein and Reisch (2014).
53 Araña and León (2013) and Eslaminassab and Ehmer (2021).
54 Farrow, Grolleau and Ibanez (2017).
55 Ramli (2021) and Mabangure and Valilai (2025).
56 Larrick and Soll (2008).
57 Varazzani, Sullivan-Paul and Tuomaila (2023).

References

Ahlgren, L. (2023) United Airlines Enlists Oscar The Grouch For Sustainability Promotion Campaign; Simple Flying; https://simpleflying.com/united-airlines-oscar-the-grouch-saf-campaign/

Araña, J.E., León, C.J. (2013) Can Defaults Save the Climate? Evidence from a Field Experiment on Carbon Offsetting Programs; in: *Environmental and Resource Economics,* Volume 54; pp. 613–626; doi:10.1007/s10640-012-9615-x

ATAG – Air Transport Action Group (2024) Aviation and Climate Change; Fact Sheet #2; December 2024; https://atag.org/media/gw5cgzzh/fact-sheet_2_aviation-and-climate-change.pdf

ATAG – Air Transport Action Group (2025) Sustainability communications in aviation: it's a matter of trust; https://atag.org/blog-posts/sustainability-communications-in-aviation-its-a-matter-of-trust

Babakhani, N., Ritchie, B.W., Dolnicar, S. (2017) Improving Carbon Offsetting Appeals in Online Airplane Ticket Purchasing: Testing New Messages, and Using New Test Methods; in: *Journal of Sustainable Tourism,* Volume 25 (7); pp. 955–969; https://www.researchgate.net/publication/309718762

Babić, D., Kalić, M., Janić, M., Dožić, S., Kukić, K. (2022) Integrated Door-to-Door Transport Services for Air Passengers: From Intermodality to Multimodality; in: *Sustainability,* Volume 14; 6503; doi:10.3390/su1411

Becker-Olsen, K.L., Cudmore, B.A., Hill, R.P. (2006) The impact of perceived corporate social responsibility on consumer behavior; in: *Journal of Business Research,* Volume 59 (1); pp. 46–53; doi:10.1016/j.jbusres.2005.01.001

Belz, F.-M., Peattie, K. (2012) *Sustainability Marketing: A Global Perspective*; 2nd edition; Wiley, West Sussex

Bernini, F., La Rosa, F. (2024) Research in the greenwashing field: concepts, theories, and potential impacts on economic and social value; in: *Journal of Management and Governance,* Volume 28 (2); pp. 405–444; doi:10.1007/s10997-023-09686-5

Bladt, D., van Capelleveen, G., Yazan, D.M. (2023) The influence of greenwashing practices on brand attitude: A multidimensional consumer analysis in Germany; in: *Business Strategy and the Environment,* Volume 33 (2); pp. 597–625; doi:10.1002/bse.3496

Buchmann, A. (2025) Willingness to pay (WTP) for leisure air travel: From WTP analysis to the conception of the 3-plane interaction model (3-PIM) as decision-making tool for airline management; in: *Journal of the Air Transport Research Society,* Volume 4; 100063; doi:10.1016/j.jatrs.2025.100063

Changi Airport (2025) Encouraging recycling in Changi Airport; https://www.changiairport.com/en/corporate/our-sustainability-efforts/environment/encouraging-recycling.html

Cocolas, N., Walters, G., Ruhanen, L., Higham, J. (2024) Air travel and persuasive climate communications; in: *Journal of Sustainable Tourism,* Volume 32 (10); pp. 2118–2140; doi:10.1080/09669582.2023.2272060

Corder, M. (2024) An Amsterdam court has ruled KLM's sustainable aviation advertising misled consumers; AP News; https://apnews.com/article/klm-greenwashing-aviation-sustainability-amsterdam-7d0e2b69099c7c9a46393a3cda44d4d8

Cordes, H., Baumeister, S., Käyrä, M. (2024) Factors influencing the willingness to pay for aviation voluntary carbon offsets: A literature review; in: *European Journal of Tourism Research,* Volume 36; 3602; doi:10.54055/ejtr.v36i.2741

Cosmo (2025) Cosmo Oil Marketing Joins Narita Airport's SAF Awareness Promotion Campaign – Supporting the Spread of Next-Generation Energy for Sustainable Future Air Travel; https://www.cosmo-energy.co.jp/en/information/press/2025/251112-01.html

Crosby, P., Thompson, T., Best, R. (2024) Air travellers' attitudes towards carbon emissions: evidence from the Google Flights interface; in: *Journal of Sustainable Tourism*, Volume 33 (12); pp. 2669–2692; doi:10.1080/09669582.2024.2412161

Dangelico, R.M., Vocalelli, D. (2017) "Green Marketing": An analysis of definitions, strategy steps, and tools through a systematic review of the literature; in: *Journal of Cleaner Production*, Volume 165, pp. 1263–1279; doi:10.1016/j.jclepro.2017.07.184

Davari, A., Strutton, D. (2014) Marketing mix strategies for closing the gap between green consumers' pro-environmental beliefs and behaviors; in: *Journal of Strategic Marketing*, Volume 22 (7); pp. 563–586; doi:10.1080/0965254X.2014.914059

de Freitas Netto, S.V., Falcão Sobral, M.F., Bezerra Ribeiro, A.R., da Luz Soares, G.R. (2020) Concepts and forms of greenwashing: a systematic review; in: *Environmental Sciences Europe*, Volume 32 (19); doi:10.1186/s12302-020-0300-3

de Mello, F.P., Macario, R. (2024) Assessing the efficacy of EU greenwashing directive: A study of European airlines' voluntary carbon offset programs; in: *Journal of the Air Transport Research Society*, Volume 3, 100028; doi:10.1016/j.jatrs.2024.100028

Delmas, M.A., Burbano, V.C. (2011) The Drivers of Greenwashing; in: *California Management Review*, Volume 54 (1); pp. 64–87; doi:10.1525/cmr.2011.54.1.64

Dobruszkes, F. (2025) Is air/high-speed rail integration the panacea to curb the impact of aviation on climate change? The case of Frankfurt Airport; in: *European Journal of Transport and Infrastructure Research*, Volume 25 (1); pp. 160–177; https://www.researchgate.net/publication/389620898

EASA – European Union Aviation Safety Agency (2024) Study on the Analysis, Prevention and Management of Air Traffic Controller Fatigue; https://www.easa.europa.eu/sites/default/files/dfu/ATCO_Fatigue_Study_Final_report_D4C1_-_V4.1_final_21_May_2024.pdf

EASA – European Union Aviation Safety Agency (2025) European Aviation Environmental Report 2025; https://www.eurocontrol.int/sites/default/files/2025-01/eurocontrol-easa-eaer-2025.pdf

EC – European Commission (2025) Twenty-one European airlines agree to modify their practices regarding environmental claims following dialogue with the Commission and national consumer protection authorities; https://ec.europa.eu/commission/presscorner/detail/en/ip_25_2608

Eslaminassab, A., Ehmer, H. (2021) Nudging Air Travellers towards Carbon Offsetting: the Impact of an Opt-Out Default on the Offsetting Rate; in: *Transportation Research Procedia*, Volume 59; pp. 271–280; doi:10.1016/j.trpro.2021.11.119

Farrow, K., Grolleau, G., Ibanez, L. (2017) Social Norms and Pro-environmental Behavior: A Review of the Evidence; in: *Ecological Economics*, Volume 140; pp. 1–13; doi:10.1016/j.ecolecon.2017.04.017

Font, X., Elgammal, I., Lamond, I. (2017) Greenhushing: the deliberate under communicating of sustainability practices by tourism businesses; in: *Journal of Sustainable Tourism*, Volume 25 (7); pp. 1007–1023; doi:10.1080/09669582.2016.1158829

Gössling, S., Dolnicar, S. (2022) A review of air travel behavior and climate change; in: *WIREs Climate Change*, Volume 14 (1); doi:10.1002/wcc.802

Greenpeace (2022) Busted: 5 tricks the aviation industry plays on us to appear green; https://www.greenpeace.org/international/story/54040/busted-5-tricks-the-aviation-industry-plays-on-us-to-appear-green/

GRI/SASB (2021) A Practical Guide to Sustainability Reporting Using GRI and SASB Standards; https://www.globalreporting.org/media/mlkjpn1i/gri-sasb-joint-publication-april-2021.pdf

Groening, C., Sarkis, J., Zhu, Q. (2018) Green marketing consumer-level theory review: A compendium of applied theories and further research directions; in: *Journal of Cleaner Production*, Volume 172; pp. 1848–1866; doi:10.1016/j.jclepro.2017.12.002

Gudmundsdottir, S., Sigurjonsson, T.O. (2024) A Need for Standardized Approaches to Manage Sustainability Strategically; in: *Sustainability*, Volume 16; 2319; doi:10.3390/su16062319

Hartmann, P., Apaolaza-Ibáñez, H. (2012) Consumer attitude and purchase intention toward green energy brands: The roles of psychological benefits and environmental concern; in: *Journal of Business Research*, Volume 65 (9); pp. 1254–1263; doi:10.1016/j.jbusres.2011.11.001

Hensher, D.A., Nelson, J.D. (2025) Do integrated mobility services have a future? The neglected role of non-mobility service providers: Challenges, and opportunities to extract sustainable transport outcomes; in: *Transport Policy*, Volume 163; pp. 348–357; https://www.sciencedirect.com/science/article/pii/S0967070X25000356

Hollenstein, L., Wittmer, A. (2024) Does low-carbon nudging work for travelers of the young generation? Nudging customers to reduce personal air travel within Europe – An experiment with Swiss Generation Y and Z travelers; in: *Journal of the Air Transport Research Society*, Volume 2; 100015; doi:10.1016/j.jatrs.2024.100015

Hung, C.-Z., Chang, T.-W. (2023) Have I purchased the right product? Consumer behavior under corporate greenwash behavior; in: *Journal of Consumer Behavior*, Volume 23; pp. 1102–1113; doi:10.1002/cb.2263

Hyams, K., Fawcett, T. (2013) The Ethics of Carbon Offsetting; in: *Wiley interdisciplinary reviews: Climate Change*, Volume 4 (2); pp. 91–98; doi:10.1002/wcc.207

ICMA – International Capital Market Association (2023) Sustainability-Linked Bond Principles – Voluntary Process Guidelines; https://www.icmagroup.org/assets/documents/Sustainable-finance/2023-updates/Sustainability-Linked-Bond-Principles-June-2023-220623.pdf

Iken, N., Aguessy, F.-X. (2022) Calculating the Greenhouse Gas Emissions of Flights: A Comparative Study of Existing Protocols; in: *IOP Conference Series Earth and Environmental Science*, January 2022, Volume 952; 012002; doi:10.1088/1755-1315/952/1/012002

Karhunmaa, K., Salmivaara, S., Varho. V., Virtanen, M.J., Eerikäinen, T., Vainio, A. (2023) Settling an unsettled phenomenon: citizens' views and understandings of voluntary carbon offsetting in Finland; in: *Journal of Environmental Policy and Planning*, Volume 25 (5); pp. 524–538; doi:10.1080/1523908X.2023.2211539

KLM (2025) SAF: alternative aviation fuel; https://www.klm.de/en/information/sustainability/saf

Kortsch, T., Händeler, P. (2024) Explaining sustainable purchase behavior in online flight booking - Combining value-belief-norm model and theory of planned behavior; in: *Gruppe. Interaktion. Organisation. Zeitschrift für Angewandte Organisationspsychologie*, Volume 55 (2); pp. 127–140; doi:10.1007/s11612-024-00735-7

Kotler, P., Armstrong, G., Opresnik, M.O. (2024): *Principles of Marketing*; 19th edition; Pearson, London

Kurek, B., Górowski, I. (2025) Willingness to Pay for Green Energy: Exploring Generation Z Perspectives; in: *Sustainability*, Volume 17 (17); 7953; doi:10.3390/su17177953

Larrick, R.P., Soll, J.B. (2008) The MPG Illusion; in: *Science*, Volume 320 (5883); pp. 1593–1594; doi:10.1126/science.1154983

Leonidou, C.N., Katsikeas, C.S., Morgan, N.A. (2013) "Greening" the marketing mix: do firms do it and does it pay off?; in: *Journal of the Academy of Marketing Science*, Volume 41 (2); pp. 151–170; doi:10.1007/s11747-012-0317-2

Li, D.C., Beck, M.J., Merkert, R. (2026) Willingness to pay for voluntary airline carbon offsets and sustainable aviation fuel; in: *Transportation Research Part A*, Volume 204; 104824; doi:10.1016/j.tra.2025.104824

Life in Norway (2025) This is Oslo's Green Airport City; https://www.lifeinnorway.net/oslo-airport-city/

Lufthansa (2024) Now available worldwide: more sustainable flying with Green Fares; https://newsroom.lufthansagroup.com/en/now-available-worldwide-more-sustainable-flying-with-green-fares/

Mabangure, T., Valilai, O.F. (2025) Investigating consumer behaviour and environmental sustainability for the impact of CO2 emission disclosure on German online product purchases; in: *Discover Sustainability*, Volume 6 (900); pp. 1–26; doi:10.1007/s43621-025-01924-0

Mahesh, S., Calvert, S.C. (2025) Decarbonizing airport access: A review of landside transport sustainability; in: *Transportation Research Part D*, Volume 140; 104625; https://www.researchgate.net/publication/389474961

Martin-Domingo, L.; Efthymiou, M.; Mota, M.M. (2025) Airline sustainability reporting in Europe: Progress, compliance and challenges; in: *Environmental and Sustainability Indicators*, Volume 28; 101008; doi:10.1016/j.indic.2025.101008

Munich Airport (2025) Climate and environmental protection; https://www.munich-airport.com/climate-and-environmental-protection-264103

Neureiter, A., Hirsch, M., Matthes, J., Naderer, B. (2024) Sustainable Flying? The Effects of Greenwashed Claims in Airline Advertising on Perceived Greenwashing, Brand Outcomes, and Attitudes Toward Flying; in: *Environmental Communication*, Volume 18 (7); pp. 873–890; doi:10.1080/17524032.2024.2327060

OECD – Organisation for Economic Co-operation and Development (2024) Bridging Talent Shortages in Tech: Skills-first Hiring, Micro-credentials and Inclusive Outreach, Getting Skills Right; https://www.oecd.org/en/publications/bridging-talent-shortages-in-tech_f35da44f-en.html

Ottman, J.A. (2011) *The New Rules of Green Marketing. Strategies, Tools, and Inspiration for Sustainable Branding*; Greenleaf Publishing Limited, Sheffield

Patino-Rataza, H., Suau-Sanchez, P. (2025) What drives passengers towards sustainable aviation? A segmentation study of travel behaviour and environmental concern in Spain; in: *Case Studies on Transport Policy*, Volume 21; 101539; doi:10.1016/j.cstp.2025.101539

Peattie, K., Crane, A. (2005) Green marketing: Legend, myth, farce or prophesy?; in: *Qualitative Market Research: An International Journal*, Volume 8 (4); pp. 357–370; doi:10.1108/13522750510619733

Peattie, K., Belz, F.-M. (2010) Sustainability Marketing: An Innovative Conception of Marketing; in: *Thexis*, Volume 27; pp. 8–15; doi:10.1007/s11621-010-0085-7

Pigou, A.C. (1932) *The Economics of Welfare*; 4th edition; Macmillan, London

Ramli, U. (2021) Social Norms Based Eco-Feedback for Household Water Consumption; in: *Sustainability*, Volume 13 (5); 2796; doi:10.3390/su13052796

Reddy, K.P., Chandu, V., Srilakshmi, S., Thagaram, E., Sahyaja, Ch., Osei, B. (2023) Consumers perception on green marketing towards eco-friendly fast moving consumer goods; in: *International Journal of Engineering Business Management*, Volume 15; pp. 1–14; doi:10.1177/18479790231170962

Ritchie, B.W., Kemperman, A., Dolnicar, S. (2021) Which types of product attributes lead to aviation voluntary carbon offsetting among air passengers?; in: *Tourism Management*, Volume 85; 104276; doi:10.1016/j.tourman.2020.104276

Rogachuk, B.E., Prigmore, S.M., Ogbaga, C.C., Okolie, J.A. (2025) Public Perception and Awareness of Sustainable Aviation Fuel in South Central United States; in: *Sustainability*, Volume 17 (9); 4019; doi:10.3390/su17094019

Schomakers, E.-M., Engelmann, L., Ziefle, M. (2024) Diversity in the acceptance of sustainable aviation fuels: Uncovering varying motivational patterns; in: *Fuel Communications*, Volume 20; 100129; doi:10.1016/j.jfueco.2024.100129

Sharma, S. (2025) Clearing the skies: Environmental challenges of aviation; Kogod School of Business; https://kogod.american.edu/news/clearing-the-skies-environmental-challenges-of-aviation

Singapore Airlines (2025) Singapore Airlines Group Continues Progress On Decarbonisation Journey with Neste and World Energy Agreements; https://www.singaporeair.com/en_UK/de/corporate/newsroom/press-release/2025/april-may/singapore-airlines-group-continues-progress-on-decarbonisation-j/

Sohail, S. (2017) Green marketing strategies: how do they influence consumer-based brand equity?; in: *Journal for Global Business Advancement*, Volume 10 (3); https://www.researchgate.net/publication/317419085

T&E – Transport & Environment (2025) Down to earth; https://www.transportenvironment.org/uploads/files/TE_Down_to_Earth_report.pdf

Thaler, R.H., Sunstein, C.R. (2008) *Nudge: Improving Decisions About Health, Wealth and Happiness*; Yale University Press, New Haven

The Guardian (2023) Air France, Lufthansa and Etihad ads banned for misleading environmental claims; https://www.theguardian.com/media/2023/dec/06/air-france-lufthansa-etihad-ads-banned-misleading-claims-advertising-environmental

Thepchalerma, T., Pinsuwan, S. (2025) CEO voices on sustainable aviation: An analysis of environmental communication in the airline industry; in: *Green Technologies and Sustainability*; Volume 3; 100194; https://www.sciencedirect.com/science/article/pii/S2949736125000284

Tseng, S-C., Hung, S.-W. (2013) A framework identifying the gaps between customers' expectations and their perception in green products; in: *Journal of Cleaner Production*, Volume 59; pp. 174–184; doi:10.1016/j.jclepro.2013.06.050

Tsoi, B.H.N., Chung, Liu, J.CE. (2025) Assessing airline communication for voluntary carbon offsets; in: *npj Sustainable Mobility and Transport*, Volume 2 (10); doi:10.1038/s44333-025-00027-7

Tu, J.-C., Cui, Y., Liu, L., Yang, C. (2024) Perceived Greenwashing and Its Impact on the Green Image of Brands; in: *Sustainability*, Volume 16 (20); 9009; doi:10.3390/su16209009

United (2023) United Names Oscar the Grouch as First Chief Trash Officer; https://united.mediaroom.com/2023-03-02-United-Names-Oscar-the-Grouch-as-First-Chief-Trash-Officer

Varazzani, C., Sullivan-Paul, M., Tuomaila, H. (2023) Behavioural Science for Sustainable Tourism - Insights and policy considerations for greener tourism; OECD Working Papers on Public Governance; https://www.oecd.org/content/dam/oecd/en/publications/reports/2023/06/behavioural-science-for-sustainable-tourism_75f42dd5/c2ec4fcf-en.pdf

Vieira J., Castro, S.L., Souza, A.S. (2023) Psychological barriers moderate the attitude-behavior gap for climate change; in: *PLoS ONE* Volume 18 (7); e0287404; doi:10.1371/journal.pone.0287404

Wendt, N., Kipperberg, G., Lindhjem, H. (2024) Consumer willingness to pay for emission reduction in air travel: A meta-analysis; in: *Transportation Research Part D*; Volume 134; 104347; doi:10.1016/j.trd.2024.104347

Wirtz, B.W. (2024) *Multichannel Marketing, Strategy – Design, Digital Technology*; Springer Gabler, Wiesbaden

Xu, M., Tse, Y.K., Geng, R., Liu, Z., Potter, A. (2025) Greenwashing and market value of firms: An empirical study; in: *International Journal of Production Economics*, Volume 284; 109606; doi:10.1016/j.ijpe.2025.109606

Zou, L., Lee, K.K., Waguespack, B. (2025) CO_2 emissions and airfare in the U.S. domestic air travel market; in: *Transportation Research Part E*, Volume 203; 104382; doi:10.1016/j.tre.2025.104382

Zurich Airport (2025a) Zurich Airport tests solar fences to boost renewable energy production; https://newsroom.flughafen-zuerich.ch/en/zurich-airport-tests-solar-fences-to-boost-renewable-energy-production/

Zurich Airport (2025b) Subglacial groove – a natural heat and cold storage; https://www.flughafen-zuerich.ch/en/company/flughafen-zuerich/airport-development/current-construction-projects/rinne

10 Actionable Recommendations

Introduction

There is a growing consensus that sustainability in aviation is not a marginal or isolated issue but a core strategic challenge arising from structural characteristics of the industry and growing societal expectations. Aviation is characterised by high capital intensity and long asset lifetimes, which result in technological lock-ins and limit the speed of transformation. Investment decisions made today shape emission trajectories for decades. While technological and operational efficiency gains have reduced emissions per passenger-kilometre, these improvements are partially offset by rebound effects, as efficiency gains lower operating costs and often airfares, thereby stimulating demand growth. As explained in Chapter 2, this leads to a risk of rising absolute emissions despite technological advances. In addition, aviation remains strongly dependent on fossil energy, as scalable technological alternatives are currently unavailable or economically not competitive. Incoherent regulatory frameworks further exacerbate this challenge by hindering consistent strategies and effective transformation processes. Furthermore, Chapter 5 has highlighted the fragmentation of international regulation, particularly between ambitious European frameworks and more lenient regimes elsewhere. As in other industries, this results in distortions of competition and carbon leakage, a risk that is structurally rooted in the historically fragmented regulation of international aviation under the Chicago Convention and is only partially addressed by ICAO mechanisms (see also Chapter 5). This fragmentation reinforces the strategic uncertainty highlighted in Chapter 2, where regulation functions simultaneously as a catalyst for sustainability and a constraint on coherent long-term planning for globally operating aviation players. At the same time, stakeholder pressure from investors, customers, employees and civil society has intensified, making sustainability performance a determinant of access to capital, legitimacy and long-term resilience.

Sustainability strategies in aviation must explicitly emphasise the economic and social benefits in order to maintain social acceptance and political legitimacy. Against this background, the purpose of this chapter is to

DOI: 10.4324/9781003608172-10

derive evidence-based, yet practice-oriented recommendations for key decision makers in the aviation system. Rather than reiterating analytical findings, the chapter translates the insights of Chapters 2–9 into actionable guidance tailored to the specific roles, responsibilities and constraints of different stakeholders.

The recommendations are addressed to three central groups:

- Airlines
- Airports
- Air navigation service providers (ANSPs)

The recommendations are guided by a multidimensional understanding of sustainability, reflecting ecological, social and economic objectives in line with the triple bottom line and ESG frameworks as discussed in Chapter 2. They explicitly acknowledge the existence of trade-offs, long investment cycles and international interdependencies, and are structured along three complementary levels: systemic and governance-related conditions, organisational strategy and operational implementation. While each set of recommendations can be applied within the respective organisational context, their effectiveness ultimately depends on coordinated action across the aviation system.

The recommendations therefore follow a three-level logic:

- Addressing systemic and governance-related conditions,
- Translating sustainability objectives into organisational strategies,
- Operationalising sustainability through concrete measures within the daily practices of aviation actors.

Overarching Principles

Before addressing individual stakeholder groups, several overarching principles can be derived that should guide sustainability strategies across the aviation system.

First, sustainable transformation requires systemic integration rather than isolated measures. This systemic perspective is particularly important given the strong empirical link between air connectivity and regional economic performance highlighted in Chapter 3. Measures that indiscriminately restrict connectivity may undermine employment, interests of tourism-dependent regions and social inclusion, especially where aviation represents an essential service rather than a discretionary mode of travel. As outlined in Chapter 2, isolated efficiency improvements risk to be offset by demand growth. Systemic integration is therefore essential to avoid a shift in emissions or a delay in implementing measures. Technological innovation, regulation, market mechanisms and behavioural change must be aligned to avoid unintended consequences such as rebound effects or carbon leakage. Second, given the global nature of aviation, international coordination is indispensable. Unilateral or regionally fragmented approaches undermine effectiveness, fairness and create the risk of shifting emissions rather than reducing them. Third, regarding CO_2 emissions priority should be given to actual emission reductions over compensation mechanisms. Offsetting may play a temporary role, but it cannot replace structural decarbonisation, and excessive use of offsets risks undermining credibility. Finally, sustainability strategies must explicitly take into account distributional effects and thus ultimately social acceptance. Measures that disproportionately burden specific social groups or regions may provoke political backlash and loss of legitimacy.

These principles form the normative framework for the following stakeholder-specific recommendations. Each set of recommendations is tailored entirely to the respective target group and can be applied independently. Overlaps and repetitions are therefore intentional. The recommendations are not structured hierarchically. Depending on the size of the organisation, it may be necessary to make a targeted selection to ensure effective and practicable implementation. However, the recommendations only achieve their full impact through holistic implementation.

Specific Recommendations for Airlines

Airlines are at the centre of the aviation system and bear key responsibility for its ecological transformation, social acceptance and economic resilience. As mentioned above, capital intensity, regulation and pressure from interest groups result in airline managers having only limited but strategically relevant leeway to act. Sustainability should therefore be understood not as a secondary consideration but as a core component of corporate strategy as Chapter 2 demonstrates that long-term profitability and access to capital increasingly depend on the ability of aviation players to proactively address environmental and social challenges within an ESG-oriented strategic framework.

Airlines are simultaneously:

- Emitters of CO_2 and responsible for non-CO_2 effects,
- Users of infrastructure and SAF,
- Employers with high social responsibility who are competing for talents,
- Competitive players in a global airline market with differing regional conditions.

Their decisions influence investors, customers, employees, airports and ANSPs as well as policymakers alike. Beyond their environmental footprint, airlines play a central role in shaping regional and global connectivity patterns. As outlined in Chapter 3, connectivity is a key driver of economic competitiveness, investment attractiveness and social mobility. Airline sustainability strategies should therefore consider not only emissions metrics, but also the socio-economic implications of network design, frequency reductions or market exits, particularly for regions with limited modal alternatives.

Airlines' scope for action is shaped by temporal horizons and regulatory constraints. In the short term, managerial discretion primarily lies in improving operational efficiency. Medium-term strategies enable influence over fleet composition, network design and staff planning. In the long term, airlines mainly exert influence through cooperation, interest representation and investment decisions. Across all time horizons, however, direct action is constrained by regulatory frameworks, including ICAO rules, the EU emissions trading scheme (ETS) and SAF mandates as well as labour and social legislation.

Implement Sustainability Systematically and Credibly

Aviation's environmental impacts are inherently multidimensional and cannot be reduced to a single emission metric. Climate effects, noise exposure and local air pollution differ with respect to spatial scale, temporal persistence and affected stakeholders. Consequently, sustainability management in aviation must explicitly recognise and manage trade-offs between these dimensions rather than optimising individual indicators in isolation. While scientific uncertainty persists regarding the exact magnitude of flight specific non-CO_2 effects, their cumulative climate relevance is sufficiently robust to justify precautionary integration into decision-making (see Chapter 4). From an environmental governance perspective, uncertainty therefore reinforces, rather than weakens, the rationale for early operational consideration. Noise and local air quality impacts are not adequately captured by average efficiency indicators. Noise annoyance is driven by peak events, temporal distribution and night-time exposure, while nitrogen oxides and particulate matter contribute to local health risks even where overall efficiency improves. As traffic volumes grow, these effects remain politically salient and socially contested, underscoring the limits of efficiency-oriented strategies that do not explicitly address absolute exposure levels.

It is recommended that airline managers

- not only exploit all available operational efficiency potentials in the short term but also systematically integrate environmental impacts (e.g., non-CO_2 effects) into corporate management plans in the medium term,
- explicitly integrate sustainability into their mission, vision and long-term corporate strategy,
- integrate a life cycle perspective into environmental management systems.

The implementation of sustainability represents a core managerial responsibility in the airline industry. Rather than constituting a stand-alone initiative, sustainability must be systematically integrated into both strategic and operational decision-making processes. A fundamental prerequisite is its consistent anchoring within the company's mission and vision. These normative frameworks provide strategic orientation, define priorities and communicate internally and externally that environmental responsibility is an integral component of the airline's business model.

Based on this foundation, clearly specified strategic objectives are required to translate sustainability commitments into actionable guidance. Such objectives define the contribution of key levers, including the deployment of Sustainable Aviation Fuel (SAF), long-term fleet renewal strategies and cooperation along the value chain with manufacturers, airports and air traffic control organisations. Strategic targets thus facilitate the transformation of abstract ambitions into concrete operational measures and investment decisions, strengthening managerial transparency and accountability.

At the operational level, numerous measures exist to reduce the ecological footprint of flight operations (see Chapter 6). Optimisation of flight trajectories, including lower or slower flight trajectories, can significantly reduce fuel burn and emissions, provided that safety and capacity constraints are respected. In addition, aircraft weight and loading optimisation constitutes a critical lever. The guiding principle that "every kilogram counts" highlights that even small weight reductions in cabin equipment or e.g., potable water carriage can generate substantial cumulative effects over large numbers of flights.

A further key element is the integration of sustainability indicators into cockpit and dispatch processes. Aviation sustainability is characterised by inherent trade-offs between climate impacts, noise abatement and local air quality. Making these trade-offs explicit through operational key performance indicators (KPIs) enables transparent and accountable decision-making (see Chapter 4). Trials and pilot initiatives, such as impact-optimised flight planning or contrail avoidance, provide experimental settings to evaluate innovative approaches under real-world conditions.

The benefits of a holistic implementation are multifaceted. Strategic coherence between corporate objectives and operational practices increases the effectiveness of sustainability measures, while the involvement of all management levels and operational personnel fosters internal acceptance and shared understanding. Externally, a consistent approach enhances credibility vis-à-vis regulators, investors and the public. However, conflicts with short-term economic objectives, organisational change requirements, operational complexity and dependence on external partners must be carefully managed. Efficiency gains must be embedded within appropriate governance and demand-side frameworks to avoid rebound effects and ensure absolute environmental improvements rather than impact displacement.

Adapt the Business Model to Changing Environmental Conditions

Companies differentiate between defensive (compliance) and proactive (opportunity) sustainability strategies. Airlines that use sustainability as a source of added value will be successful in the long term.

It is recommended that airline managers

- view sustainability not only as a cost and risk factor, but also as a potential source of innovation and differentiation,
- actively integrate sustainable products and services into the core business in order to position the airline as a pioneer of change in terms of long-term competitive advantage.

The development of sustainable product and service offerings is a key lever for transforming sustainability from a purely efficiency and compliance-based issue into a source of strategic value creation. For management, this means systematically integrating environmental and social aspects into the service portfolio and linking them to the company's existing core competencies. Sustainable offerings include not only "greener" products in a narrower sense, but also accompanying services, new pricing models, transparency solutions and consulting services that support customers in achieving their own sustainability goals.

One key strategic approach is to position oneself as a so-called "transition leader". This refers to a company that does not simply orient itself towards fully mature markets or clear regulatory requirements but actively helps shaping the transition to more sustainable market structures. Such positioning requires the courage to innovate, early investment and a willingness to set new standards. At the same time, it enables learning curve effects to be realised, technological and organisational competencies to be built up and the company's own brand to be linked with sustainability and responsibility. From an economic perspective, Chapter 2 demonstrates that sustainability-oriented strategies may reinforce long-term competitiveness by strengthening a firm's legitimacy, employer attractiveness and access to capital. Airlines that integrate sustainability into their business model not only respond to regulatory pressure but may also benefit from productivity gains, reputational advantages and improved stakeholder relations, thereby aligning environmental responsibility with economic performance rather than treating it as a trade-off.

The expected benefits of this strategy are manifold. Sustainable product and service offerings can be used to tap into competitive differentiation potential, especially in markets where environmental criteria are increasingly influencing purchasing decisions. Companies that credibly position themselves as pioneers can set themselves apart from their competitors and realise price premiums or intensify long-term customer relationships. In addition, a clear sustainability positioning strengthens customer and employee loyalty. Customers identify themselves more strongly with companies whose values match their own, while employees perceive meaningfulness and future prospects as important factors for motivation and loyalty. In the long term, this contributes to a more robust, sustainable business model that is better prepared for tighter regulations, resource scarcity and changing social expectations.

At the same time, this strategic orientation involves considerable risks and preconditions. A key source of uncertainty lies in the difficulty of predicting market developments and regulatory conditions. Path dependencies, for example in relation to technologies, funding mechanisms or standards, can mean that investment decisions made at an early stage prove to be inadequate. This increases the investment risk, especially if sustainable offerings involve high upfront costs and their economic viability only becomes apparent in the medium to long term.

In the short term, there may also be a decline in sales or pressure on margins, for example due to higher production costs, lower economies of scale or customers' willingness to pay still being limited. For management, this means actively managing and transparently communicating conflicts of interest between short-term financial performance and long-term strategic positioning. Another key requirement is close coordination with external stakeholders. Developing sustainable offerings often requires collaboration with suppliers, technology partners, customers, regulatory authorities or within industry initiatives. Without coordinated interfaces, common standards and reliable data flows, sustainable value chains can only be realised to a limited extent.

Overall, it is clear that sustainable product and service innovations and positioning the airline as a transition leader are less an isolated innovation project and more a comprehensive transformation task. Their success depends largely on strategic clarity, organisational learning capacity and the ability to actively shape complex stakeholder networks.

From a Sustainability Marketing perspective, the adaptation of airline business models corresponds primarily to the "Green Product" and "Green Price" dimensions of the Green Marketing Mix developed in Chapter 9. Sustainable Aviation Fuels, transparent emissions information, differentiated pricing models and sustainability-linked services should therefore not be positioned as peripheral add-ons, but as core elements of the airline's value proposition. This integration enhances both market differentiation and the perceived credibility of sustainability commitments.

Align Fleet and SAF Strategies in a Realistic and Investor-Friendly Manner

Long investment cycles, SAF shortages and high costs for SAF create an innovation dilemma. At the same time, investors are increasingly linking access to capital to credible ESG strategies. Consistent with the findings in Chapter 7, this innovation dilemma underlines that technological solutions such as SAF are necessary but insufficient, and must be embedded in broader strategic, regulatory and demand-side approaches.

It is recommended that airline managers

- align their fleet and SAF strategies in a consistent, realistic and transparent manner over the long term and actively communicate these to capital markets.

A consistent fleet and financing strategy plays a key role in the operational implementation of sustainability in capital-intensive industries such as aviation. The central task for management is to consistently align long-term investment decisions with efficiency, climate and regulatory targets and to integrate these into existing corporate management.

A key starting point is prioritising efficient aircraft types and the ongoing transition to lighter aircraft with more modern, fuel-efficient engines. Fleet modernisation has a direct impact on fuel consumption, emission intensity and noise pollution, making it one of the most effective levers for reducing the environmental footprint. In addition, retrofitting existing aircraft can deliver efficiency gains in the short term. Aerodynamic improvements such as winglets or innovative surface coatings ("sharkskin") reduce drag and help to reduce fuel consumption and emissions without requiring immediate replacement investments (see Chapter 6).

Another key component is the conclusion of long-term purchase agreements for SAF. Such off-take agreements create planning security for both sides. Airlines secure early access to scarce sustainable fuels, while producers gain investment security for the development of corresponding production capacities. On the financing side, linking capital market instruments to sustainability goals is becoming increasingly important. The issuance of so-called green bonds or sustainability-linked financing enables companies to make their environmental goals transparent and integrate them into their investment logic in a binding manner. Consistent with the logic of Green Pricing discussed in Chapter 9, sustainability-linked financing instruments translate environmental performance into economic incentives, thereby internalising external costs and strengthening the credibility of long-term decarbonisation strategies.

The expected benefits of this integrated strategy lie primarily in increased planning security, improved access to capital markets and a reduction in regulatory risks. Companies that invest in efficient technologies and use sustainable financing instruments at an early stage are better prepared for stricter environmental regulations and benefit from growing interest from sustainability-oriented investors.

At the same time, there are relevant risks and prerequisites. The economic viability of many measures depends on stable political conditions and support mechanisms. In addition, the lack of a global level playing field can lead to competitive disadvantages vis-à-vis market participants who are subject to lower regulatory requirements. Successful implementation therefore requires long-term political reliability and international coordination of climate protection measures.

Strengthen Social Sustainability and Improve Working Conditions

As demonstrated in Chapter 3, social sustainability in aviation extends beyond internal human resource considerations and directly affects safety, productivity, corporate reputation and social acceptance. Labour relations, diversity, working conditions and corporate social responsibility constitute critical elements of long-term resilience in a safety-critical and knowledge-intensive industry such as aviation.

It is recommended that airline managers

- establish social sustainability as a strategic success factor.

The operational implementation of sustainability increasingly extends beyond environmental aspects and encompasses social dimensions that are highly relevant for practitioners. In safety-critical and knowledge-intensive industries such as aviation, the design of duty and rest period schemes represents a key lever. Improved scheduling concepts that better reflect physiological and psychological load patterns contribute to higher alertness, reduced fatigue and ultimately enhanced safety and productivity. From a managerial perspective, such models require a shift from purely cost-driven rostering towards a more holistic assessment of long-term performance and risk.

Diversity programmes constitute a further important element of socially sustainable operations. Targeted initiatives to increase the participation of underrepresented groups such as women in pilot roles, engineering functions or senior management positions support a broader talent base and foster more inclusive organisational cultures. Empirical research suggests that diverse teams enhance problem-solving capabilities and decision quality, which is particularly valuable in complex operational environments. Beyond internal benefits, visible commitment to diversity strengthens the organisation's reputation and legitimacy in society. Sustainability training for employees complements these measures by embedding shared values, knowledge and competencies across the organisation. Training programmes that address ESG topics enable employees to understand how sustainability objectives relate to their daily tasks and decision-making responsibilities. This, in turn, increases engagement and ownership of sustainability initiatives.

The expected benefits of this integrated approach include improved safety outcomes, higher productivity, increased attractiveness as an employer and stronger social acceptance. However, implementation is associated with short-term cost increases and must navigate complex labour laws and collective bargaining frameworks. Successful execution therefore requires early stakeholder involvement, transparent communication and alignment with existing regulatory and contractual structures.

Strategically Professionalise Sustainability Communication

Sustainability communication is an integral component of Sustainability Marketing, particularly in emission-intensive industries such as aviation, where it actively shapes stakeholder perceptions, trust and behaviour (see Chapter 9). From an environmental economics perspective, it fulfils a corrective function by addressing market failures such as negative externalities and information asymmetries. Transparent and reliable information supports informed decisions by consumers and investors and complements regulatory instruments aimed at internalising external costs. Sustainability communication is therefore not merely reputational but a key element of effective environmental governance, while misleading or vague claims increase Greenwashing risks and undermine credibility and long-term brand equity.

It is recommended that airline managers

- make sustainability communication strictly fact based, standard compliant and verifiable,
- establish sustainability communication as a strategic marketing discipline with clear rules and governance.

As emphasised in Chapter 5, fragmented regulatory frameworks increase the risk of competitive distortions and carbon leakage, a problem further exacerbated by non-standardised or non-verifiable sustainability claims. Aligning corporate sustainability communication with recognised standards, reporting frameworks and certification schemes enhances regulatory coherence and reduces misleading signals in international markets. In line with the Green Marketing Mix outlined in Chapter 9, sustainability communication must be consistent with product characteristics, pricing, distribution and promotional practices, as inconsistencies increase reputational and regulatory risks. This requires close institutional linkage between marketing, sustainability management and operations, supported by clear governance structures, shared performance indicators and the integration of sustainability KPIs into performance management and, where appropriate, remuneration schemes. Beyond formal structures, this also entails extending the safety-related "just culture" to sustainability issues, thereby encouraging employees to identify and report inefficiencies such as excessive fuel consumption or inadequate waste handling in support of objectives like a zero-waste strategy.

A clear distinction between emission reduction and offsetting measures is equally essential. While emissions trading operates under binding caps, offsetting schemes involve higher uncertainty regarding additionality and permanence. Sustainability communication must reflect these differences to avoid overstating climate benefits and misleading claims. As shown in Chapter 9, effective communication must also account for heterogeneous stakeholder expectations, as passengers, investors, regulators, employees and the media respond to different credibility signals. Effective Sustainability Marketing therefore requires differentiated communication formats:

- Simplified, benefit-oriented narratives for passengers,
- Data-driven, comparable indicators for investors,
- Transparent methodological explanations for regulators and civil society.

In line with the analysis in Chapter 5, stakeholder groups differ not only in their information needs but also in their underlying evaluative rationalities. While consumers may respond primarily to simplified narratives and perceived environmental benefits, regulators and policymakers assess sustainability performance in terms of effectiveness, economic efficiency, distributional effects and compatibility with existing policy instruments. Sustainability communication must therefore reflect these differing logics to remain credible across stakeholder groups. Uniform sustainability messages risk either oversimplification or incomprehensibility and may thereby erode trust rather than strengthen it. Tailored communication enhances relevance, comprehension and credibility across stakeholder groups. In this context, the deliberate use of the Green Marketing Mix – encompassing product design, pricing mechanisms, distribution choices, and promotional activities (as also outlined in Chapter 9) – supports the consistent positioning of sustainability across market-facing activities.

From a policy-instrument perspective, Chapter 5 classifies information-based measures taken by policymakers as "soft" instruments that rarely achieve emission reductions on their own but play an important supporting role in enhancing the acceptance and effectiveness of stricter regulatory and market-based measures. Similarly, sustainability communication should therefore be designed as a complementary instrument that reinforces regulatory objectives rather than as a substitute for substantive environmental action. Embedding sustainability targets into budgeting and planning processes further strengthens implementation. When environmental and social objectives are considered alongside financial metrics, sustainability becomes an integral part of resource allocation and strategic trade-off decisions rather than an add-on.

The expected benefits include greater corporate credibility, reduced legal and reputational risks and long-term trust building with key stakeholders. Additional advantages arise from increased employer attractiveness and improved access to capital markets. Enhanced ESG transparency reduces information asymmetries for investors and can contribute to lower capital costs.

However, these benefits are contingent on robust implementation. Increased reporting requirements and associated costs must be managed, and inconsistent execution poses significant credibility risks. Reliable data structures, clear responsibilities and rigorous governance mechanisms are therefore essential prerequisites for success. Moreover, weak or inconsistent implementation of information-based instruments may trigger more intrusive regulatory interventions. In this sense, ineffective sustainability communication does not only pose reputational risks but may also increase regulatory pressure, as policymakers respond to perceived governance failures with stricter command-and-control measures.

Anticipate Regulatory Requirements ("Regulatory Readiness")

Environmental policy follows the principle of internalising external costs. Instruments such as an ETS, SAF mandates, noise charges and operational restrictions are being progressively tightened and combined. Airlines with purely reactive compliance bear disproportionate costs.

It is recommended that airline managers

- anticipate regulatory developments early and integrate them as a strategic variable in investment and network decisions.

Building on Chapters 5 and 7, regulatory readiness should be understood as an organisational capability that links operational environmental measures with structured monitoring, reporting and verification (MRV) systems. Its operational implementation requires systematic regulatory monitoring covering relevant policy instruments and operational constraints, including emission trading schemes, SAF regulations and noise provisions. Continuous surveillance enables early detection of regulatory developments and assessment of their operational and financial implications. In line with Chapter 6, this monitoring should explicitly include operational measures that are subject to increasing regulatory control, such as noise abatement procedures and APU substitution. Anticipating their potential transition from voluntary best practices to mandatory requirements enables airlines to proactively adjust their standard operating procedures and investment priorities. Scenario analyses are a complementary management tool that goes beyond monitoring ETS price developments or the availability of SAF and incorporates the cumulative mitigation potential of operational measures, thereby supporting informed investment and fleet planning decisions.

The expected benefits of this forward-looking approach include reduced volatility in regulatory costs, greater planning certainty and improved strategic coherence. As highlighted in Chapter 7, sustainability reporting is increasingly serving as an early warning system, as expanded MRV frameworks and disclosure requirements often precede substantial regulatory tightening. Airlines that systematically analyse new reporting requirements can therefore anticipate regulatory developments earlier than their competitors, enabling timely adaptation and competitive advantages.

Preparing for regulatory requirements involves forecasting uncertainties, additional analysis effort and coordination costs. Effective implementation requires reliable data, expertise and integrated data management structures to ensure consistency between operational data, sustainability reporting and regulatory filings. Chapters 6 and 7 also highlight limitations arising from dependence on external stakeholders such as airports and air navigation service providers, as well as potential discrepancies between reporting requirements and the pace of operational mitigation measures, which increase the risks associated with temporary compliance.

Specific Recommendations for Airports

Airports act as pivotal system integrators within the aviation ecosystem. They

- provide long-lived infrastructure that shapes operational and investment decisions over decades,
- host and manage local environmental impacts from their airline customers, including noise, NO_x, particulate matter and their own operational footprint,
- are major energy consumers with growing potential as energy producers and enablers of an energy transition,
- manage key interfaces between airlines, air navigation service providers, regulators and passengers,
- serve as the most visible touchpoint for local communities, making airports central to social acceptance and the translation of sustainability objectives into operational practice, despite their limited direct influence on aircraft-related emissions.

As outlined in Chapter 3, airports generate substantial direct, indirect, induced and catalytic economic effects, especially employment and income generation, e.g., through additional tourism revenue. Sustainability strategies at airport level must therefore balance environmental objectives with their role as regional economic anchors. Transparent communication of both positive economic contributions and environmental trade-offs is essential to maintain the social licence to operate and avoid polarised public debates.

Airports' scope for action varies across time horizons. In the short term, management attention focuses on operational measures, energy management and proactive stakeholder communication. Medium-term levers include infrastructure upgrades, the design of airport charges and incentive schemes and the decarbonisation of ground operations. In the long term, airports can shape sustainability outcomes through strategic infrastructure investments, integrated energy systems and improved intermodal connectivity.

Across all time horizons, however, airport strategies are constrained by a dense regulatory environment. National regulation, concession agreements, planning and environmental law, noise protection regimes, EU state aid rules and limited influence over airlines' fleet and network decisions restrict unilateral action. Effective airport management therefore requires aligning sustainability ambitions with regulatory compliance and collaborative governance approaches.

Embed Sustainability as a Core Element of Airport Strategy and Governance

Airports' long investment cycles and strong local visibility require early and credible integration of sustainability into strategic decision-making. Fragmented or purely operational approaches risk lock-ins and loss of legitimacy.

It is recommended that airport managers

- embed sustainability explicitly in mission, vision and long-term infrastructure planning and anchor it at board level,
- integrate ESG targets into master planning, investment decisions and management incentives, supported by transparent stakeholder engagement to manage trade-offs and build legitimacy.

The operational implementation of sustainability at airports requires a structured and integrated management approach that embeds environmental and social objectives into core decision-making processes. A key lever is the systematic integration of sustainability targets into airport master plans and long-term investment decisions, ensuring that infrastructure development, capacity planning and decarbonisation pathways are aligned from the outset. Participation within established frameworks such as the Airport Carbon Accreditation (ACA) programme (see Chapter 7), and its continuous expansion to higher accreditation levels, further supports standardisation, benchmarking and credibility in climate action.

Internally, effective implementation is strengthened by linking management objectives and remuneration schemes to clearly defined ESG targets. This alignment helps translating strategic ambitions into operational accountability and signals that sustainability is a core management priority rather than a peripheral concern. At the same time, sustainability should not be treated solely as an internal target system. Airports increasingly act as visible platforms for external value positioning, communicating their role in enabling more sustainable aviation ecosystems. In this context, emphasis should be placed on credibility, transparency and consistency of actions, rather than on the mere communication of aspirational targets.

Given their strong local embeddedness, airports also benefit from institutionalised stakeholder engagement. The establishment of stakeholder advisory boards enables structured dialogue, early identification of conflicts and the incorporation of local perspectives into planning and operational decisions.

The expected benefits of this approach include greater strategic coherence across functions, improved social acceptance and legitimacy as well as reduced regulatory and reputational risks. However, these benefits are contingent on managing inherent trade-offs, particularly between capacity growth and noise or climate objectives. Transparent governance structures and robust, reliable data are therefore essential prerequisites to balance competing objectives and support informed, credible decision-making.

Use Pricing and Other Incentives to Steer Airline Decisions Towards Sustainability

Although airports do not exercise direct control over aircraft emissions, they can significantly influence airline decisions through the strategic design of charges, fees and incentive mechanisms.

It is recommended that airport managers

- systematically use airport charges and fees to incentivise less noise, lower local emissions and cleaner operations,
- design airport charges not only in a technically differentiated manner, but also in a communicative and customer-oriented way in order to increase acceptance among airlines and reduce resistance.

From an operational perspective, implementation may include differentiated landing and noise charges based on noise exposure as well as proxies for emissions such as NO_x and CO_2. Additional incentives can be offered for the deployment of less noisy aircraft, the use of low-emission ground service equipment and the adoption of Sustainable Aviation Fuels. Central to the effectiveness of these measures is the establishment of fee structures that are transparent, predictable and stable over time, thereby enabling airlines to incorporate environmental considerations into long-term fleet and operational planning.

The expected benefits of this approach include behavioural change without the need for direct regulatory intervention, as well as a closer alignment of local environmental objectives with economic performance at the airport level. However, several risks and prerequisites must be considered. These include the potential for airline resistance or traffic leakage to competing airports, as well as the requirement for regulatory approval and a charging design that is demonstrably non-discriminatory and compliant with applicable legal frameworks.

Decarbonise Airport Energy Systems and Infrastructure

Airports are among the most energy-intensive transport infrastructures and therefore represent critical leverage points for achieving emission reductions beyond aircraft-related operations.

It is recommended that airport managers

- prioritise energy efficiency and renewable energy supply as key pillars of their sustainability strategy.

Operational implementation should combine the deployment of on-site renewable energy generation such as photovoltaic systems and geothermal installations with off-site renewable electricity procurement through power purchase agreements. In parallel, comprehensive energy efficiency measures are essential, including the replacement of conventional lighting with LED systems, improvements in building insulation and the application of smart building management systems to optimise energy consumption in real time. Further emission reductions can be achieved through the electrification of ground handling equipment and airport vehicle fleets, complemented by the provision of electric vehicle charging infrastructure on the landside to support employees, passengers and service providers.

The expected benefits of this approach include substantial direct reductions in CO_2 emissions, lower long-term energy costs and reduced exposure to energy price volatility. Moreover, enhanced energy autonomy increases the resilience of airport operations to external supply disruptions. Beyond environmental performance, such infrastructure investments can generate reputational advantages and strengthen the competitive positioning of airports in an increasingly sustainability-oriented aviation market. However, these measures are associated with high upfront capital requirements and may be constrained by grid capacity limitations and complex approval procedures, which must be addressed through careful planning and stakeholder coordination.

Enable Low-Emission Ground and Aircraft Operations

Operational processes at airports exert a substantial influence on (sustainable) fuel consumption, local air quality and noise exposure. Consequently, airports represent an important leverage point for reducing emissions beyond airborne operations. The development of SAF infrastructure enables the storage, blending and distribution of Sustainable Aviation Fuels and represents a critical measure for reducing aviation-related CO_2 emissions.

It is recommended that airport managers

- proactively invest in scalable SAF infrastructure and partnerships to ensure future readiness and regulatory compliance,
- actively enable and promote low-emission operational procedures in close cooperation with airlines and Ground Handling Service Providers.

Operational implementation should focus on systematically replacing the use of Auxiliary Power Units (APUs) by providing fixed power connections on the ground and pre-conditioned air at aircraft parking stands, thereby reducing both fuel consumption and local pollutant emissions. Further emission reductions can be achieved through the use of CO_2-free ground equipment, including electric or hydrogen-powered vehicles and machinery. In addition, efficient runway design and operational support for taxiing with one engine can reduce fuel consumption and noise during ground movement phases. Optimising de-icing and anti-icing measures, including the recovery and recycling of glycol-based fluids, also contributes to reducing environmental impacts without compromising safety standards. The use of SAF requires the integration of SAF into existing fuel systems, adapted logistics, quality assurance and close cooperation with fuel suppliers, refuelling company and airlines.

The expected benefits of low-emission ground and aircraft operations are manifold. Airports can significantly reduce local air pollutant emissions and noise, directly benefiting surrounding communities and airport staff. Improved air quality and lower noise levels contribute to better working conditions for ground staff. At the same time, more efficient operations can increase the overall efficiency of the airport and reduce operating costs for airlines and service providers. Key benefits of providing SAF infrastructure include regulatory compliance and enhanced reputation.

However, several risks and prerequisites must be addressed. Successful implementation requires close coordination with airlines and ground handlers to ensure procedural alignment and operational acceptance. In addition, investments in infrastructure, equipment and staff training are necessary to enable low-emission operations at scale and to ensure their safe and effective integration into daily airport activities.

Strengthen Transparency, Communication and Local Acceptance

Noise exposure and local air pollutant emissions remain decisive factors shaping public acceptance of airport operations. Technical mitigation measures alone are often insufficient to address societal concerns.

It is recommended that airport managers

- professionalise sustainability and noise communication and proactively involve local stakeholders in a structured and continuous manner.

Operational implementation should include more comprehensive and transparent noise reporting that goes beyond average indicators and explicitly addresses peak noise events and temporal distribution. Such reporting provides a more realistic representation of community exposure and supports informed public discourse. In addition, airport operators should clearly communicate the trade-offs between operational efficiency, capacity and environmental protection, as well as the mitigation measures implemented to manage these trade-offs. Communication strategies should be differentiated and customer focused, addressing the specific information needs of residents, local authorities, policymakers and other stakeholder groups. Structured and institutionalised dialogue formats, such as regular community forums or consultative committees, can facilitate constructive engagement and mutual understanding.

The expected benefits of strengthened transparency and communication include increased trust, legitimacy and social acceptance of airport operations. By improving stakeholder understanding of operational constraints and environmental measures, airports can reduce conflict potential and mitigate legal and planning risks associated with opposition and litigation. Enhanced transparency also supports more stable long-term development perspectives.

However, several risks and prerequisites must be considered. Proactive communication raises stakeholder expectations and may increase scrutiny of airport performance. To maintain credibility, airport operators must therefore ensure the availability of consistent, accurate and robust data. Without high-quality and verifiable information, communication efforts risk undermining trust rather than strengthening it. Long-term commitment and internal alignment across airport organisations are essential to sustain effective and credible stakeholder engagement.

Develop Climate Resilience and Adaptive Capacity

Climate change is increasingly affecting airport operations through rising temperatures, more frequent heatwaves, heavy rainfall and other extreme weather events.

It is recommended that airport managers

- systematically integrate climate adaptation considerations into infrastructure planning, asset management and enterprise risk management frameworks,
- treat resilience as a core strategic objective rather than a purely technical concern.

Operational implementation should include the development and deployment of heat-resilient runways, taxiways and pavements designed to withstand higher surface temperatures and prolonged thermal stress. Enhanced drainage systems and flood protection measures are essential to manage more intense precipitation events and reduce the risk of surface water disruption. In addition, climate-related risks should be explicitly integrated into asset management processes, including life cycle planning, maintenance strategies and investment prioritisation. Beyond technical measures, climate resilience should be positioned as a visible and integral component of the airport's sustainability narrative, reinforcing the link between adaptation, long-term operational reliability and societal responsibility.

The expected benefits of systematically building climate resilience include higher operational robustness, reduced vulnerability to weather-related disruptions and lower long-term costs associated with emergency repairs and unplanned downtime. Proactive adaptation can also protect asset value and support more reliable service provision under increasingly volatile climate conditions.

However, several risks and prerequisites must be addressed. Uncertainty in climate projections complicates the definition of appropriate design standards and investment horizons. Furthermore, long planning, approval processes can delay the implementation of adaptation measures. Addressing these challenges requires flexible design approaches, scenario-based planning and early engagement with regulators and stakeholders to ensure that resilience investments are timely, proportionate and aligned with long-term development strategies.

Clear Allocation of Responsibilities within Airports

A transparent and well-documented allocation of responsibilities within airport organisations is a fundamental prerequisite for achieving sustainability objectives, in line with frameworks such as the European Green Deal, national climate targets and relevant Airport Council International (ACI) policies. A shared understanding of roles, accountabilities and interfaces across all organisational levels is essential to ensure regulatory compliance, consistent implementation and the systematic integration of sustainability principles into airport planning, operations and infrastructure development.

It is recommended that airport managers

- clearly define, document and communicate roles, responsibilities and accountabilities related to environmental management and sustainability across strategic governance, infrastructure planning and operational functions,
- ensure that these responsibilities are supported by structured, role-based training and competency management for management and staff, explicitly integrating climate protection, environmental performance, resource efficiency, noise and emission mitigation as well as energy management.

From an organisational and operational perspective, coherent implementation of airport sustainability objectives requires a clear distinction between strategic governance functions, infrastructure and capacity planning and day-to-day operational activities. Strategic airport management should be accountable for setting long-term sustainability targets, ensuring compliance with environmental regulation and integrating climate and environmental considerations into investment decisions, master planning and performance management systems. This includes defining measurable targets related to emissions reduction, energy efficiency, noise management and resource use, and embedding these targets into corporate governance and reporting structures. Infrastructure planning and development functions should be responsible for translating strategic sustainability objectives into environmentally optimised airport layouts, building designs and operational concepts. This includes the integration of sustainable infrastructure principles such as energy-efficient terminals, low-emission ground operations, electrification of ground service equipment, optimised taxiway and apron layouts and measures to mitigate local noise and air quality impacts while maintaining safety and capacity requirements. Operational airport functions should be responsible for the environmentally optimised execution of daily activities, including aircraft ground handling, airside and landside traffic flows, energy use, waste management and collaboration with airlines and service providers. This includes the application of sustainable operational practices such as reduced engine-on times, efficient stand allocation, optimised turnaround processes and the use of data-driven decision-making to minimise emissions and resource consumption.

Clearly defined interfaces, coordination mechanisms and escalation procedures between strategic, planning and operational functions are essential to support the overall system optimisation. These interfaces should be consistently reflected in organisational documentation, operational procedures and training programmes to ensure the uniform application of sustainability principles across all airport functions.

The expected benefits of a clear allocation of responsibilities, supported by appropriate training and competency assurance, include improved accountability, enhanced environmental performance, more efficient sustainability-oriented decision-making and the timely implementation of technological, procedural and organisational measures. Transparent responsibility structures also support performance monitoring and continuous improvement by enabling a clear attribution of environmental and sustainability outcomes within integrated management systems.

However, several risks and prerequisites must be addressed. Organisational silos, overlapping mandates, insufficient sustainability-related competencies and resistance to change may undermine effective implementation. These challenges require robust governance arrangements, clear management accountability and structured change management processes, including early stakeholder involvement, clear internal communication and continuous competency development with a strong focus on sustainability performance.

Specific Recommendations for ANSPs

Air navigation service providers (ANSPs) play a central role in shaping the environmental and operational performance of the aviation system. In particular, they

- operate at the interface between airlines, airports and regulators, directly influencing flight efficiency,
- implement regulatory frameworks that define safety, capacity, cost and performance targets, while retaining operational discretion,
- shape environmental and efficiency outcomes through airspace configuration and the design of flight procedures,
- influence operational behaviour via the provision and use of flight data,
- have limited but relevant scope to affect incentives through charging schemes.

Effectively leveraging this room for manoeuvre is critical to improving environmental performance and operational efficiency without compromising system resilience.

Embed Sustainability as a Core Element of ANSP Strategy and Governance

ANSPs occupy a system-critical position between regulators, airlines and airports. Their operational and planning decisions have immediate impacts on fuel efficiency, emissions and network performance. Fragmented or purely technical approaches to environmental measures risk underutilising available operational levers and may lead to inconsistent implementation across airspace boundaries.

It is recommended that ANSP managers

- actively embeds sustainability and environmental performance as explicit objectives in corporate strategy, performance plans and governance structures.

If sustainability is not systematically embedded in ANSP strategy and governance, significant operational risks may arise. Isolated or ad hoc environmental initiatives can lead to inconsistent implementation across airspace boundaries, reducing overall network efficiency and predictability. The lack of a strategic focus increases the risk that available operational levers such as flight path optimisation, balancing demand and capacity or cross-border coordination will not be sufficiently exploited. Moreover, without clear governance structures, sustainability considerations may be subordinated to short-term operational priorities, resulting in limited and non-scalable impacts.

By integrating sustainability as an explicit strategic goal, ANSPs can leverage their system-critical role to achieve measurable environmental and operational benefits. Strong governance frameworks support coherent decision-making at the planning and operational levels, enabling reductions in fuel consumption and emissions while maintaining safety and capacity. A strategic approach also improves coordination with airlines, airports and regulatory authorities, contributing to improved network performance and resilience. Furthermore, clear sustainability goals strengthen institutional credibility and support compliance with changing regulatory and societal expectations.

Potential risks include conflicts between environmental targets and capacity or cost efficiency targets. Successful implementation therefore requires clearly defined responsibilities, robust performance indicators, reliable data availability and sufficient technical as well as organisational capacities. Strong commitment from management and good coordination between the various stakeholders are essential prerequisites for effectively implementing sustainability goals in operational practice.

Cross-Border Implementation and Harmonisation of Free Route Airspace

The cross-border implementation and harmonisation of Free Route Airspace (FRA) is a key enabler for more efficient and environmentally sustainable air traffic management at the network level.

It is recommended that ANSP managers

- liaise with their respective governments,
- pursue coordinated approaches to FRA deployment that extend beyond national airspace boundaries in order to unlock the full potential of trajectory-based operations.

Operational implementation of harmonised FRA requires close coordination in airspace design, including the alignment of entry and exit points, route availability and vertical limits. In addition, procedures for flight planning, air traffic flow and capacity management and tactical control must be harmonised to ensure consistency across borders. Interoperable ATM systems and data-sharing mechanisms are essential to support seamless trajectory optimisation across multiple control areas, enabling flights to operate along fuel- and time-efficient trajectories rather than being constrained by sector-specific route structures.

The expected benefits of cross-border FRA implementation are substantial. Optimised flight trajectories can significantly reduce flight time and, thus, fuel consumption, GHG emissions and non-CO_2 effects, while also enhancing predictability for airspace users. At the network level, harmonised FRA improves capacity utilisation and strengthens overall network efficiency and resilience.

However, several risks and prerequisites must be addressed to ensure successful implementation. Regulatory fragmentation, differences in national operational practices and varying levels of technical maturity among ATM systems can impede harmonisation efforts. Overcoming these challenges requires strong international governance frameworks, sustained cooperation between states and service providers and a high level of mutual trust. Clear governance structures, shared performance objectives and transparent decision-making processes are critical to achieving durable and effective cross-border FRA harmonisation.

Standardised Deployment of Continuous Climb and Descent Operations

The standardised deployment of Continuous Climb Operations (CCO) and Continuous Descent Operations (CDO) constitutes an effective operational measure to enhance the environmental and operational performance of air traffic management.

It is recommended that ANSP managers

- liaise with airports and airlines,
- promote the systematic and harmonised implementation of CCO and CDO across terminal airspace, moving beyond isolated or situational applications toward routine use under suitable traffic conditions.

Operational implementation requires the development of standardised procedures that are consistently applied across airspace volumes and airports.

This includes the integration of CCO/CDO into terminal area design, the alignment of arrival and departure procedures, and the use of compatible arrival and departure management tools that support trajectory-based operations. Comprehensive controller training is essential to ensure a shared understanding of CCO/CDO concepts, operational constraints and decision-making principles. In addition, close coordination with airports and airlines is necessary to ensure that flight management systems and operational practices are compatible with continuous profiles.

The expected benefits of standardised CCO/CDO deployment are well documented, as outlined in Chapter 6. By minimising level-offs and reducing unnecessary thrust changes, continuous profiles lead to lower fuel consumption, reduced GHG and local air pollutant emissions and decreased noise exposure in communities surrounding airports.

However, several risks and prerequisites must be considered. High traffic density, complex airspace structures and mixed aircraft equipment can limit the feasibility of continuous operations, particularly during peak periods. Without effective traffic flow and capacity management, the environmental benefits of CCO/CDO may not be fully realised. Addressing these challenges requires careful airspace design, demand-capacity balancing and the progressive integration of advanced ATM tools to support continuous operations under increasingly complex conditions.

Data-Driven Planning and Congestion Forecasting

Data-driven planning and congestion forecasting constitute essential enablers for improving the efficiency and environmental performance of ATM.

It is recommended that ANSP managers

- liaise with airlines and other ANSPs,
- systematically leverage advanced data analytics to support strategic and tactical decision-making.

Operational implementation relies on the use of advanced analytics, predictive modelling and shared data platforms that integrate information from flight plans, surveillance systems, meteorological services and (in Europe) network management functions. These tools enable more precise forecasts of airway and sector congestion, allowing for earlier and more targeted interventions such as flow management measures or trajectory adjustments. Improved predictability also supports more accurate fuel planning by airlines, reducing the need to carry excessive contingency fuel to account for uncertainty in delays or rerouting.

The expected benefits of data-driven congestion forecasting are significant. Lower contingency fuel carriage directly reduces fuel burn, GHG emissions and operating costs, while simultaneously improving payload efficiency and aircraft performance. At the network level, enhanced predictability contributes to smoother traffic flows, better capacity utilisation and improved punctuality, benefiting both operators and passengers.

However, the effectiveness of data-driven planning depends on several critical prerequisites. High data quality, consistent standardisation and timely data sharing across organisational and national boundaries are essential to ensure reliable forecasts and robust decision-making. In addition, the increasing reliance on digital platforms heightens the importance of cyber-security and data protection. Without adequate safeguards and governance frameworks, risks related to data integrity, confidentiality and system resilience could undermine trust and limit the adoption of data-driven approaches in air traffic management.

Enhanced Meteorological Services and ISSR Prediction

Enhanced meteorological services, including improved prediction of ice-supersaturated regions (ISSRs), represent a promising instrument for reducing the non-CO_2 climate impacts of aviation.

It is recommended that ANSP managers

- cooperate with meteorological authorities and airlines to systematically integrate advanced meteorological information into air traffic management and flight operations in order to enable informed decisions aimed at contrail mitigation.

Operational implementation relies on the development of high-resolution weather models and the integration of real-time atmospheric data into ATM and flight planning systems. Accurate identification and forecasting of ISSRs allow for tactical altitude or lateral adjustments that can avoid persistent contrail formation while maintaining safe and efficient operations. To be operationally viable, meteorological outputs must be translated into clear, actionable guidance and embedded within existing decision-support tools for pilots and air traffic controllers.

The expected benefits of enhanced ISSR prediction are substantial. Targeted contrail avoidance has the potential to significantly reduce the overall climate impact of aviation, often with minimal additional fuel burn or operational penalties when applied selectively. By focusing on a limited number of high-impact flights or airspace regions, these measures can deliver disproportionate climate benefits compared to their operational cost, thereby complementing longer-term technological and fuel-based mitigation strategies.

However, several risks and prerequisites must be addressed. Forecast uncertainty remains a key challenge, as inaccuracies in ISSR prediction can reduce effectiveness or lead to unnecessary deviations. In addition, the increased cognitive workload for pilots and controllers associated with dynamic trajectory adjustments may limit practical applicability. These challenges underscore the need for a high degree of automation, robust decision-support systems and clear operational procedures. Strong coordination between meteorological service providers and ATM stakeholders is essential to ensure reliability, trust and acceptance of ISSR-based operational measures.

Green Navigation Charges Based on Environmental Performance

The introduction of environmental navigation charges based on performance represents a market-based instrument for promoting more sustainable air transport. There are different options with regard to terminal and en-route charges.

It is recommended that ANSP managers

- liaise with regulatory bodies,
- integrate environmental criteria into navigation charging schemes in order to internalise external environmental costs and influence aircraft operators' decisions without relying solely on prescriptive regulation.

Operational implementation requires the development and application of transparent, objective and verifiable metrics that directly link navigation charges to aircraft noise levels, local air pollutant emissions and climate-related impacts. Such metrics may be based on established proxies for NO_x emissions, aircraft certification data or standardised performance indicators. Clear communication of the underlying methodology and predictable charging structures are essential to enable airlines to integrate environmental cost signals into fleet planning and day-to-day decision-making.

The expected benefits of green navigation charges include the creation of market-based incentives for fleet modernisation, the uptake of cleaner aircraft technologies and the adoption of more environmentally efficient operational practices. By rewarding superior environmental performance, these charging schemes can accelerate innovation and support the alignment of economic efficiency with environmental objectives across the air traffic management (ATM) system.

However, the effectiveness and acceptance of green navigation charges depend on several critical prerequisites. Robust and harmonised indicators are necessary to ensure consistency across airspace users and to avoid competitive distortions. In addition, legal alignment with existing charging regulations and international frameworks is essential to guarantee non-discriminatory application and regulatory certainty. Without such safeguards, there is a risk of reduced acceptance among stakeholders and challenges to the legitimacy of environmental differentiation in navigation charging systems.

Strengthen Social Sustainability and Improve Working Conditions within ANSPs

Social sustainability within ANSPs extends beyond technological and operational measures and depends significantly on human performance, organisational resilience and working conditions. Within a safety-critical ATM environment, environmental performance is closely linked to the ability of controllers, engineers and managers to consistently apply fuel-efficient and environmentally optimised operational concepts. Social sustainability, human factors and environmental outcomes are therefore interdependent and should be addressed in an integrated manner within the Safety Management System (SMS), for example in line with EASA ATM regulatory requirements.

It is recommended that ANSPs

- establish environmental sustainability and climate performance as explicit objectives within their SMS and performance management systems, supported by working conditions and human-factor-oriented measures that enable environmentally optimised ATM operations.

From an operational and organisational perspective, the effective implementation of environmental sustainability within ATM requires embedding environmental considerations into SMS processes such as hazard identification, safety risk management and safety assurance. In this context, the design of working conditions and workload management is a critical enabler. Excessive workload, fatigue and time pressure can limit the ability of operational staff to apply fuel-efficient procedures, optimise trajectories or support dynamic airspace use. Conversely, sustainable working conditions support higher situational awareness, better decision-making and the consistent application of environmentally optimised operational concepts. ANSPs should therefore ensure that rostering, staffing and capacity management practices support the reliable delivery of environmental performance objectives. This includes aligning operational staffing levels with the complexity of traffic flows and environmental optimisation tasks, particularly during peak periods or when implementing new green ATM concepts. Environmental sustainability should be explicitly considered as part of operational risk assessments within the SMS.

To support safety assurance and performance monitoring, ANSPs should define and track environment-related KPIs that are linked to human performance and operational decision-making, such as:

- CO_2 emissions per traffic unit within the ANSP's area of responsibility,
- Rate of application of fuel-efficient procedures (e.g., continuous climb and descent operations),
- Vertical and horizontal flight inefficiencies,
- Frequency of tactical interventions that increase flight distance or level flight during climb or descent due to capacity or workload constraints.

Training and competency management are essential to ensure that environmental sustainability objectives are effectively implemented in day-to-day ATM operations. Training programmes for management and operational staff should integrate environmental awareness, ATM principles aiming at fuel efficiency and the interaction between human performance, workload and environmental outcomes. This includes scenario-based training on environmentally optimised traffic management, trade-offs between capacity, safety and emissions as well as the use of decision-support tools that enable greener operations. Relevant KPIs include training coverage on environmental topics, demonstrated competency in green ATM procedures and trends in environment-related performance indicators following training implementation.

The expected benefits of this integrated approach include improved environmental performance, reduced fuel burn and emissions, enhanced safety margins and increased operational predictability. By aligning working conditions, human factors and environmental objectives within the SMS, ANSPs can more effectively contribute to sustainable aviation while maintaining safety and capacity. Transparent governance structures and measurable KPIs support continuous improvement and regulatory oversight.

However, several risks and prerequisites must be addressed. Environmental optimisation may introduce additional complexity into ATM operations and decision-making, particularly under high-traffic or degraded conditions. Insufficient staffing, inadequate training or weak integration of environmental objectives into the SMS can undermine effectiveness. Successful implementation therefore requires strong leadership commitment, clear accountability for environmental performance, early involvement of operational staff and continuous monitoring of environmental KPIs, ensuring that sustainability objectives are achieved without compromising safety or resilience.

Clear Allocation of Responsibilities within ATM

A clear, transparent and well-documented allocation of responsibilities within the ATM system is a fundamental prerequisite for the achievement of performance and sustainability objectives, e.g., in line with the Single European Sky (SES) Performance Scheme, the European Green Deal and the ICAO Global Air Navigation Plan (GANP). A shared understanding of roles, accountabilities and interfaces across all organisational levels is essential to ensure consistent implementation, regulatory compliance and the systematic integration of sustainable aviation principles into ATM planning and operations.

It is recommended that ANSP managers

- clearly define, document and communicate roles, responsibilities and accountabilities across the ATM system, covering strategic governance, airspace design and planning and operational service provision,
- ensure that these responsibilities are supported by structured, role-based training and competency management for management and staff, explicitly integrating sustainable aviation objectives, environmental performance, climate impact mitigation and fuel-efficient ATM operations, in accordance with applicable national, regional and ICAO requirements.

From an operational and organisational perspective, coherent implementation in support of sustainable aviation can be achieved by clearly distinguishing responsibilities between strategic ATM governance functions, airspace design and planning functions and operational ATM service provision. Strategic management should be accountable for setting long-term performance and sustainability objectives, ensuring compliance with the regulatory framework and defining performance targets, e.g., under the SES Performance Scheme, including environmental and climate-related targets. In this context, strategic governance should ensure that sustainability considerations are systematically embedded in ATM decision-making, investment planning and operational priorities. Airspace design and planning functions should be responsible for translating these strategic sustainability objectives into operationally feasible airspace structures, procedures and concepts of operations that minimise environmental impact, support fuel-efficient trajectories and balance safety, capacity and environmental performance. This includes the integration of sustainable airspace design principles, such as trajectory optimisation, reduced fragmentation and the mitigation of noise and emissions impacts, while maintaining required safety levels. Operational ATM service provision should be responsible for real-time traffic management, tactical flow and capacity adjustments, and the safe, efficient, resilient and environmentally optimised execution of agreed operational concepts. This includes the application of sustainable operational practices, such as fuel-efficient trajectory management, dynamic use of airspace and environmentally optimised tactical decision-making.

Clearly defined interfaces, coordination arrangements and escalation mechanisms between these functions are necessary to support effective decision-making, safety assurance, environmental performance management and overall system optimisation. These interfaces should be consistently reflected in organisational documentation, operational procedures and training programmes to ensure the effective and uniform application of sustainable aviation principles across all ATM functions.

The expected benefits of a clear allocation of responsibilities, supported by appropriate training and competency assurance, include improved accountability, enhanced safety and environmental performance, more efficient and sustainability-oriented decision-making processes, and the timely implementation of technological, procedural and organisational changes that support sustainable aviation. Well-qualified management and staff contribute to reduced coordination inefficiencies, more predictable and environmentally optimised operational outcomes and improved overall ATM system performance. In addition, transparent responsibility structures support performance monitoring and continuous improvement by enabling clear attribution of safety, performance and sustainability outcomes within integrated safety and performance management systems.

However, several risks and prerequisites must be addressed. Organisational silos, overlapping mandates, insufficient sustainability-related competencies and resistance to change may undermine the effectiveness of responsibility frameworks. These challenges require robust governance arrangements, clear management accountability and structured change and safety management processes, e.g., in line with ICAO SMS principles. This includes early stakeholder involvement, clear internal communication and continuous training and competency development for management and staff, with a specific focus on sustainability and environmental performance, to ensure a resilient, sustainable and compliant implementation.

General Recommendations for Collaboration

Building on the theoretical foundations developed in Chapter 8, cooperation between airports, airlines and air navigation service providers (ANSPs) should be understood not as a downstream coordination mechanism, but as an integral part of a sustainability-oriented corporate strategy. Chapter 8 argues that sustainability is increasingly shaping missions and visions, thereby fundamentally changing the strategic framework for action of organisations. Value creation in aviation therefore no longer takes place in isolation within individual organisations, but is embedded in complex, interdependent ecosystems of operational decision-makers and regulatory institutions. In this context, cooperation becomes a strategic prerequisite for achieving environmental objectives under given operational and economic constraints.

Chapter 8 also shows that sustainability-oriented strategies promote learning processes and innovation capabilities within organisations. These effects can be systematically reinforced through inter-organisational cooperation. The exchange of knowledge, data and experience across organisational boundaries enables collective learning processes, shared experimentation spaces and the distribution of technological and financial risks. Especially in a capital-intensive and highly path-dependent industry such as aviation, cooperative innovation reduces the inhibition threshold for investments in sustainable technologies and accelerates their diffusion along the value chain.

Another vital point is the holistic consideration of sustainability challenges throughout the entire life cycle. Emissions and environmental impacts arise in aviation from raw material extraction to manufacturing and operation to decommissioning and recycling. Since no single decision-maker has all the relevant levers at its disposal, fragmented individual strategies are not only inefficient but also carry the risk of shifting the burden between actors. Cooperation is therefore structurally necessary in order to address systemic effects and avoid unintended rebound or displacement effects.

It is essential to distinguish between defensive, compliance-oriented and opportunity-oriented sustainability strategies and their application to potential forms of cooperation. While defensive cooperation primarily aims at cost sharing, regulatory protection and risk minimisation, opportunity-oriented cooperation opens up potential for joint value creation, innovation and strategic differentiation.

There are already a number of operational measures that can be implemented with limited infrastructure investment but with a significant impact.

Cooperation between airports and ANSPs plays a central role in mitigating aircraft noise. Procedures such as Continuous Descent Approach (CDA), displaced thresholds and steeper approach paths reduce noise exposure for communities by keeping aircraft at higher altitudes for longer periods. Preferential runway use and Noise Abatement Departure Procedures (NADPs) further support local noise management strategies, particularly during night-time operations or in noise-sensitive areas. These measures require harmonised procedure design, precise navigation and consistent operational compliance. Joint initiatives between airlines and ANSPs focus increasingly on data-driven and trajectory-based operations. Enhanced air-ground data exchange enables real-time decision support, allowing pilots and onboard systems to access more accurate and timely weather information. This supports contrail-avoiding flight trajectories through optimised vertical and lateral planning, contributing to climate impact mitigation. Improved data sharing also enables better fuel planning and operational predictability, while allowing cleaner aircraft configurations during approach, such as reduced thrust and optimised flap settings. Collaboration between airports and airlines can significantly reduce ground emissions. Supporting single-engine taxi operations lowers fuel burn and local emissions, while the long-term introduction of electric taxiing systems or operational towing offers further potential. Beyond aircraft operations, enabling intermodal "door-to-door" passenger services can shift surface access away from private car use toward more sustainable transport modes. In line with the socioeconomic analysis in Chapter 3, coordinated action between airlines, airports and ANSPs is particularly critical for safeguarding essential connectivity in peripheral regions, tourism-dependent economies and areas with limited modal alternatives. Collaborative approaches enable efficiency gains and emission reductions without disproportionately undermining regional development, social inclusion or access to essential services such as medical transport and emergency response.

Consistent with insights presented in Chapter 8, collaborative sustainability strategies are subject to structural limitations arising from long investment cycles, technological lock-ins and asymmetric stakeholder interests. Effective collaboration therefore requires clear governance structures, defined roles and mechanisms for conflict resolution. Without strategic alignment and mutual commitment, collaborative initiatives risk stagnation or lowest-common-denominator outcomes rather than transformative impact.

In summary, effective cooperation between airports, airlines and ANSPs is a key strategic prerequisite for reducing the environmental impact of aviation while ensuring operational efficiency. Building on Chapter 8, it is clear that cooperation should not be seen as a reactive measure to meet regulatory requirements, but as an active component of sustainability-oriented business models. By pooling knowledge, resources and innovative capacity, opportunity-oriented cooperation enables systemic sustainability challenges to be overcome that no single player can solve alone. Cooperation thus becomes an essential lever for sustainable value creation in aviation.

Outlook

Aviation managers cannot steer sustainable change on their own, but they can play a decisive role in shaping it. Those who are successful are those who manage environmental constraints, social responsibility and economic realities in an integrated manner, reflecting the integrated, management-oriented understanding of sustainability developed in this book, which views aviation not as an isolated sector but as a socio-technical system embedded in global economic and regulatory structures. Aviation managers who anticipate regulatory developments, leverage operational excellence, deploy technologies strategically and cooperate systemically ensure not only compliance with regulations but also competitiveness in an increasingly regulated aviation system. Beyond technological and regulatory measures, behavioural economic instruments discussed in Chapter 9 provide underutilised leverage for sustainability implementation. Nudges, social norm communication and well-designed feedback mechanisms can support environmentally preferable choices by passengers, airlines and operational staff, particularly where price or regulatory instruments face political or social limits. While such tools cannot substitute for structural emission reductions, they enhance the effectiveness and acceptance of sustainability strategies when embedded within a coherent Sustainability Marketing framework. As emphasised in Chapter 3, aviation's contribution to economic development, social participation and regional cohesion remains substantial. Sustainable transformation will therefore succeed only if environmental objectives are embedded within a broader socioeconomic narrative that recognises aviation not merely as an emitter, but as a critical infrastructure underpinning modern economies and societies.

11 Conclusions

Over the last decades, sustainability in aviation has developed from a marginal, predominantly voluntary commitment to a central framework condition for corporate action. For management, this represents a fundamental paradigm shift: sustainability is no longer to be understood as an isolated environmental or communication project, but as a cross-cutting management task that affects strategy, investment decisions, operational processes, governance structures and stakeholder relationships in equal measure. In an industry characterised by high capital intensity, long asset life cycles, global interdependencies and high public visibility, this development is unfolding with particular dynamics and complexity.

Sustainability in aviation is not a new concept. Environmental issues – particularly aircraft noise – have been taken into account since the 1950s, long before sustainability became a widely used term in management and politics. These early efforts, often driven by pressure from the communities living near airports and by new legal requirements, also laid the foundation for a tradition of voluntary commitments within the aviation industry. Over time, this tradition has expanded beyond noise to include pollutants and greenhouse gases, occupational health and safety, and selected social aspects. These voluntary commitments by the industry have contributed to significant improvements. Especially in recent years, however, the scope, depth and systemic relevance of sustainability have increased significantly and now exceed what can be effectively achieved through voluntary approaches alone. What once worked through voluntary improvement mechanisms is now proving insufficient.

This development is reflected in the increasing integration of environmental, social and governance (ESG) aspects into social expectations, financial market necessities and regulatory requirements. Sustainability is thus becoming a decisive factor in corporate legitimacy ("licence to operate") and long-term value creation. For management, this creates a tension between short-term economic targets and long-term transformation requirements. Investments in lower-emission aircraft, digital efficiency improvements or improved working conditions are often associated with considerable investment, long payback periods and uncertainties. In a highly competitive

DOI: 10.4324/9781003608172-11

market environment with high price sensitivity, these measures lead to a short-term strain on profitability.

However, this book shows that this apparent dichotomy between economic efficiency and sustainability depends largely on the perspective taken. In the short term, sustainability measures can indeed conflict with profit targets, especially if regulatory requirements vary from region to region or customers' willingness to pay remains limited. In the long term, however, sustainable strategies offer considerable substantial economic potential. This includes the reduction of regulatory risks, improved access to capital, greater attractiveness as an employer, stronger customer loyalty and increased organisational resilience to external shocks. Sustainability thus functions less as a cost factor and more as an instrument of strategic risk management and securing the future.

At the beginning of 2025, the sustainability landscape in European aviation entered the next phase of implementation. The most important regulatory frameworks – including regulations for Sustainable Aviation Fuel (SAF), emission trading, reporting and selected social aspects – have largely moved from conception to implementation or have become even more rigid. For aviation industry players operating in Europe, the management challenge has shifted from defining goals to achieving results in an efficient manner. While this increases complexity, it also provides greater predictability and rewards early investment in governance, sustainability marketing and organisational capabilities. At the same time, sustainability regulation is increasingly diffusing beyond Europe. Countries outside the EU and the USA have begun to introduce sustainability-related aviation requirements, often inspired by European mandates. This trend is further underscored by China's introduction of a CO_2 pricing mechanism, signalling a growing willingness among major non-Western economies to use market-based instruments to reduce emissions. Although regulatory fragmentation remains a challenge, this trend reduces the risk that European aviation operates under uniquely restrictive conditions. Instead, sustainability is gradually becoming a global license-to-operate issue. For globally active aviation players, early alignment with robust sustainability practices lowers long-term compliance costs and supports consistent operating models across regions. In contrast, the situation in the USA in 2025 has been marked by reprioritisation and political uncertainty. While sustainability remains relevant in regulatory agencies, financial markets and parts of corporate strategy, its political momentum has weakened, and its future trajectory is less predictable. This ambiguity does not diminish the strategic importance of sustainability for aviation firms. On the contrary, it underscores the value of proactive, self-driven approaches that reduce dependence on shifting political signals and preserve international credibility.

Taken together, these developments indicate that aviation has entered a post-hype phase of sustainability rather than a post-sustainability era. Public discourse may be less idealistic and less emotionally charged than in previous years, but the structural drivers remain firmly in place. Climate constraints, social expectations and governance requirements continue to shape the industry's operating environment, albeit in a more pragmatic and execution-oriented manner.

Against this backdrop, aviation players face two fundamental strategic options. The first is a compliance-oriented approach that focuses on meeting mandatory legal requirements at minimum cost. This strategy emphasises risk avoidance and operational efficiency but tends to treat sustainability as an external constraint rather than an internal driver of value creation. While such an approach may be sufficient in the short term, it risks leaving firms vulnerable to future regulatory tightening, reputational challenges and missed innovation opportunities. The second option is a proactive, opportunity-driven approach that integrates sustainability into the core business model. This strategy goes beyond compliance and seeks to leverage sustainability as a source of innovation, differentiation and new revenue streams. Examples include the development of Sustainable Aviation Fuel (SAF), noise-optimised flight procedures and socially responsible employment models. While this path requires greater strategic ambition and organisational transformation, it offers the potential to reconcile sustainability objectives with long-term profitability.

From a management and practical perspective, this means that sustainability should not be viewed as a binary choice between regulatory compliance and competitiveness, but rather as a strategic continuum. Companies must consciously choose their position on this continuum and implement it consistently. Key success factors include integrating sustainability into corporate strategy, linking investment decisions to long-term regulatory development paths and establishing robust governance and reporting systems. Consistent and transparent communication that builds trust without resorting to symbolic or inconsistent signals is equally crucial.

For aviation managers, sustainability is no longer a question of intent, ambition or narrative. It is a question of execution. Leadership teams must make explicit, binding choices on how sustainability is embedded in fleet renewal, fuel sourcing, capital allocation, organisational incentives and external positioning. Deferring decisions, relying on minimum compliance or outsourcing responsibility to specialised units will increase strategic exposure and future adjustment costs. The decisive factor is not whether sustainability objectives exist, but whether they are translated into governance, investment discipline and measurable operational outcomes. In an industry defined by long asset cycles and irreversible capital commitments, today's decisions will lock in competitive positions for decades. The ability to act early, coherently and credibly on sustainability has therefore become a defining test of leadership in aviation.

Index

For Product Safety Concerns and Information please contact our EU representative GPSR@taylorandfrancis.com
Taylor & Francis Verlag GmbH, Kaufingerstraße 24, 80331 München, Germany

www.ingramcontent.com/pod-product-compliance
Ingram Content Group UK Ltd.
Pitfield, Milton Keynes, MK11 3LW, UK
UKHW020147200926
475702UK00016B/2582

9781041000907